FREQUENCY ENGINEERING IN MOBILE RADIO BANDS

William M Pannell C.Eng., M.I.E.R.E

Published by
Granta Technical Editions
in association with
Pye Telecommunications Ltd

Published by Granta Technical Editions,
7 Brooklands Avenue, Cambridge in association with
Pye Telecommunications Ltd, a member of the Pye
of Cambridge Group.

First published in Great Britain 1979

ISBN 0 906782 00 7

Printed and bound in Great Britain by
William Clowes (Beccles) Limited, Beccles and London

Contents

List of Appendices

Author's Note

The rapidly growing field of Land Mobile Radio is already severely taxing the skills of frequency planning authorities throughout the world. With the possibility that the World Administrative Radio Conference being held in September of this year may be unable to provide sufficient additional spectrum to cope with the envisaged growth up until the end of the century, it is obvious that the best use must be made of the bands finally assigned.

Frequency Engineering in Mobile Radio Bands examines the aspects which must be considered carefully if maximum benefit from the portions so allotted is to be assured. Optimum frequency bands, the use of single and two frequency operation and the types of systems to be encountered are but three of the sections covering the general problem. Amongst other chapters are those considering the best frequency block sizes, block spacings and the subdivision of the blocks into channels both for general use and for use on communal sites. Emphasis throughout is on reducing all forms of interference to the minimum possible levels and to maximise spectrum usage by so doing.

Channel occupancy is also examined in detail, whilst the need for adequate record keeping, whether by manual or computer means, has not been forgotten.

Many of the aspects of frequency planning have common factors and parameters, and rather than encouraging constant reference to other chapters, the author has included, at the rear of the book, appendices on the relevant subjects. Each appendix, taken from original papers and 'in-house' engineering notes, is complete in itself and such subjects as frequency re-use, intermodulation, channel occupancy calculations, system configurations, antenna parameters, transmitter noise, etc, are examined.

The book is intended mainly to assist those Authorities who are in the early stages of frequency planning rather than the established bodies with many years experience. Nevertheless, it is believed that all readers will find points of interest.

Preface

In the Autumn of 1979, the International Telecommunications Union (I.T.U.) will be holding a World Administrative Radio Conference in Geneva at which all frequency bands, presently ranging from 10 kHz to 40 GHz, will be reviewed and re-allocated in accordance with an agreed world-wide pattern.

Decisions made at this Conference may remain current until late in the century when the next Conference is likely to be convened. During the two intervening decades, any major decisions involving changes concerning all participants may require the calling of an Extraordinary Administrative Radio Conference (such as the one in 1963 to resolve Space Communication frequencies).

The last W.A.R. Conference was held in Geneva in 1959.

At the 1979 W.A.R. Conference strong pressure will undoubtedly be made to increase the amounts of spectrum available for both private and public land mobile purposes. This pressure is made necessary by the fast growing needs of all types of user of this service and the resultant impact on the economic growth of all countries has been highlighted in many areas of the World.

Continued growth of land mobile communications will therefore depend in the first case upon the granting of suitable segments of the VHF and UHF spectrum by the W.A.R. Conference. However, this must be accompanied by an ordered pattern of use. Without such discipline, the available spectrum will not be utilised to the full, resulting in excessive loading of the channels and high levels of spectrum pollution.

In some countries with much of the existing mobile radio spectrum allocated, the pattern is already showing up problems such as electromagnetic incompatibility and overloaded channel capacity. These possibly stem from such shortcomings as record deficiency, initial planning errors, unforeseen growth and other planning problems. The 1979 World Administrative Radio Conference comes at an opportune moment to correct these deficiencies and to provide a basis for a sound future planning strategy.

By applying the principles given in the later sections of this guide to any new bands made available by the 1979 W.A.R.C., the allocations in existing bands can be run down until a point is reached where the principles can be, in turn, applied to those parts of the spectrum. Thus, over a few years it should be possible to re-allocate all channels in a manner calculated to relieve congestion, pollution etc., and to achieve the maximum use of frequency spectrum.

There are many criteria. Block layout, bandwidth, mode of operation, occupancy, channel re-use, and communal site frequency planning, are but a few. This guide is intended to highlight such points whilst suggesting methods for the maximum utilisation of the allocated spectrum, both in new and existing segments of the available bands.

1. The Frequency Planning Tree

1.1 The first echelon – Worldwide/Regional

The World Administrative Radio Conferences held under the auspices of the I.T.U. are concerned with the basic division on a worldwide basis of the frequency spectrum from 10 kHz to over 40 GHz (Ref. 1).

To achieve realistic sub-division of the spectrum and to permit some local variations within the plan, the world has been considered as three Regions (shown in Fig. 1.1).

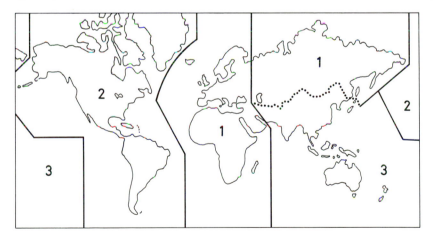

Fig. 1.1 World Regions for Frequency Planning

Region 1   Europe, Africa, Middle East (all areas west of the Arabian Gulf/Iran) and the U.S.S.R. A demarcation line extends between Greenland and Iceland and continues south, splitting the Atlantic Ocean roughly into two, whilst the eastern boundary extends due South from the Arabian Gulf through the Indian Ocean.

Region 2   North and South America, Caribbean and the Eastern part of the Pacific Ocean.

Region 3   South East Asia, Australasia, Western and Central Pacific.

It will be noted that, in practically all cases, the boundaries are formed by geographical features, such as sea, high mountain ranges, or uninhabited remote areas. Thus each Region can be considered individually insofar as user blocks are concerned, although international or regional agreement is necessary in worldwide essential services, such as aeronautical, marine and satellite.

In 1959, Conference decisions, together with the revisions needed for satellite use and agreed at the Extraordinary Administrative Radio Conference in 1963, have been the basis for all regional allocations until the next W.A.R. Conference in 1979.

In the 1959 plan, some 37 categories of use of the spectrum have been defined and each has been allocated sections of the total spectrum in amounts based on their effective possible use from 1959–1979 as discussed and agreed at the 1959 Conference.

It is the purpose of the 1979 Conference to re-appraise the decisions made at the previous Conference and to modify the allocations in accordance with the recommendations advanced by the administrations represented. These recommendations will be based on the decisions reached within the various countries at meetings held prior to the W.A.R.C. It is therefore essential that the delegates are fully aware of the requirements within their individual administrations and that full cognisance is taken of both the technical and economic needs for the next 20 years. Subsequently, major changes could well be difficult to implement without an Extraordinary Conference.

Examining the categories put forward in 1959 and showing potential users of the various spectrum segments, a great deal of sharing by different users of spectrum segments can be seen. With internationally essential services, however, sharing is often not attempted, but where geographical boundaries can be utilised, some sharing is a major contribution to an extended use of the spectrum.

Sharing of land mobile radio with, for instance, TV services, has been undertaken in certain areas of the world. The success of this attempt has shown that, provided physical spacing and geographical boundaries are employed to the full, a real saving of valuable spectrum is achievable.

In some large countries, areas remote from the sea or navigable rivers often make limited use of the International marine band for strictly local services. Interference control however is rigidly enforced in these cases.

Within the decisions reached at the W.A.R.C. – and shown by the numbered footnotes in the published tables – agreement must also be reached as to the precautions necessary and the limitations to be imposed in the use of the segments, particularly where services between neighbouring countries overlap.

1.2   The second echelon – National

Having obtained agreement as to the available segments within a region and having due regard to the restrictions, the administration in the countries concerned must now consider the best method of allotting parts of these segments to the different categories of users within their individual countries.

At this stage, and within the scope of this guide, we are only concerned with the segments affecting land mobile radio. Therefore the different categories under this general heading (public, private, security etc.) must be considered individually, noting their relative priorities and needs as an indication of the amounts to be allocated to each. The opportunity must also be taken at this level to provide the necessary guard bands or physical spacing between users and between base transmitters and receivers.

1.3 The third echelon – area allocation of discrete channels

This important stage is necessarily piecemeal and its success or failure depends not only upon the care in assigning correctly related channels in any area, but also upon the attention paid to the arrangement of the blocks at the second echelon stage. Either can cause spectrum pollution, with the incorrect layout of blocks causing a countrywide problem and incorrect channel assignment a local problem. It is essential to keep adequate and correct records of all allocations at this stage.

1.4 The fourth echelon – system allocation of discrete channels

This final stage is in many ways identical with the previous echelon but with a greater interference potential.

It concerns closely knit systems as would be the case on a communal site where correct frequency assignments based on minimum interference are needed. This is to eliminate both local and distant spectrum pollution. Both waste spectrum by preventing its effective use over considerable areas.

The frequencies allocated to specified systems must be compatible with those systems.

This guide is concerned with Echelons 2, 3 and 4 where related to land mobile radio. The subsequent sections deal in greater detail with the points to be observed for maximum frequency usage and minimum spectrum pollution.

2. Basic Rules of Land Mobile Radio Frequency Planning

It cannot be over-emphasised that the key to maximum use of the frequency spectrum segments allocated to any country depends largely upon the *initial* steps taken by that country.

However, frequency planning involves many steps and a brief outline of the more important points follows. Later sections in this guide discuss the principles in greater depth.

2.1 Common international boundaries

Those countries having common borders with other countries must work in close collaboration with their neighbours, adopting similar techniques and obeying the same basic rules.

This is particularly so where countries have extended common borders and where one or more of the countries involved is small in area. In these cases a common policy or even total co-operation may be the only method of avoiding mutual interference and achieving maximum spectrum usage.

Where protection is afforded by the size of the countries involved, common policy along the border need only extend far enough into each country to minimise local interference.

2.2 Adequate international boundaries

Where international boundaries are adequate as, for example, in neighbouring countries separated by mountains, desert, sea, or forest, then the frequency planning requirements within each segment become a country's internal concern. However, it is to the advantage of all countries to follow the same basic rules of good engineering practice. This will ease any problems which can occur under unusual propagation conditions and enable equipment parameters to be standardised to the advantage of all users.

2.3 Choice of the operational mode for land mobile use

Here we consider either single frequency or two frequency simplex. The problems associated with the incorrect choice are explained at length later. Here it is sufficient to state that two frequency simplex has been recognised as the mode enabling the optimum use to be made of the available spectrum whilst ensuring minimum spectrum pollution. Although appearing at first sight to be uneconomical, requiring two frequencies per channel instead of one, major advantages exist and multiple systems can be sited under conditions unsuitable for single frequency operation.

It should be noted that all countries using single frequency simplex operation at VHF have, without exception, resorted to two frequency operation on allocating channels in the UHF bands. Problems arising in urban areas or crowded sites have been exacerbated by the use of single frequency allocations. Clearly, it is essential to follow a rigorous procedure from the beginning to make maximum use of the spectrum.

Single frequency simplex should not be totally rejected however – it has some essential applications – but the location of the channels using this mode must follow a recognised pattern if spectrum pollution is to be avoided.

2.4 The use of discrete blocks of frequencies

This is essential for full advantage to be taken of two frequency operation. By this method all base transmitters can be allocated frequencies in discrete blocks suitably spaced from similar blocks in which all base receiver frequencies are located. This enables the co-siting of multiple systems in which transmitter and receiver frequencies are respectively separated to give the best transmitter to receiver isolation. Mutual interference is thus minimised.

Separate blocks would normally be used for public and private systems.

It should be appreciated that the occasional single frequency channel required can be suitably located in its own block relative to all other blocks so as to provide an adequate degree of transmitter to receiver isolation.

Radio links should be allocated channels in blocks reserved for point to point use only and not intermixed with mobile channels. Two frequency operation is essential if shortcomings such as lockout and repeater site incompatibility are to be avoided. As with mobile channels, neighbouring countries must be considered if there are common boundaries.

Fortunately the lowest point to point link frequency bands are often allocated, both nationally and internationally, within discrete segments of the UHF band, generally removed from land mobile operations. Therefore, provided such segments are sub-divided carefully, mutual interaction between links and mobile services can generally be avoided.

2.5 Channel Bandwidth

Channel bandwidth depends on current technology. Throughout the world, the transmitted bandwidth and the acceptance bandwidth of receivers have steadily decreased as techniques have improved. Channels 100 kHz wide in the 1950's have systematically been reduced and 12½ kHz is normal in the VHF band in a number of countries.

In the UHF bands, at this stage of development, 25 kHz is normal although here again a reduction to 12½ kHz will eventually occur.

However, reduction in bandwidth has been based on splitting or halving techniques. Therefore, whether 25 kHz or 12½ kHz channels are initially allocated, is a decision dictated by likely future growth and the available spectrum. Further splitting will, of course, depend upon future technological innovation.

Frequency stability plays a large part in the reduction of channel bandwidths and Appendix 15 deals in considerable detail with optimum frequency control of transmitters and receivers.

2.6   Intermodulation

Individual channels should be located on the basis of optimum local freedom from on-channel intermodulation. This means that receivers within an agreed distance of a number of transmitters should not be subject to intermodulation products falling on the channel to which any receiver is tuned. This infers a pattern of allocation determined by a number of basic factors. This procedure is discussed at length later in Section 12 and in Appendices 6, 7, 8, 9 and 10.

2.7   Communal Site usage of Channels

Allocations here must follow a similar but more stringent pattern to that mentioned in 2.6. Multiple usage, particularly in urban areas, infers a greater overall density of transmitters and therefore full agreement as to the location of the multiple site is often essential in optimising frequency allocation and minimising spectrum pollution.

2.8   Channel re-use

In countries having an area sufficiently large to permit re-use of VHF and UHF channels, the effective usage of the spectrum is improved. However, care must be exercised in determining the distance between sites using the same channel. Where a common user is concerned, re-use of a channel may be purely a question of operational procedure and control, and distance protection is not normally necessary. Where different users are concerned, however, the re-use distance must follow specified rules if interference is to be avoided and maximum channel usage is to be obtained. The subject is dealt with in detail in Appendices 1, 2, 3 and 4.

2.9   Channel occupancy

The loading of each channel during its busiest period should not exceed a recognised recommended level if advantages are to be derived from the use of land mobile radio. Excessive occupancy means a poorer service, longer waiting times before access and generation of a greater level of unwanted spurious products in closely knit areas. An Appendix (5) to this guide deals with this problem in detail.

2.10   Sharing of a channel in a single area by a number of small users

Where a channel is occupied by a single small user, greater utilisation of the channel can be achieved by several such users occupying that channel in the same area on a time sharing basis.

This is a complex subject if the sharing is to be satisfactory and a later section of this guide together with Appendix 1 indicates the necessary measures.

2.11 Point to Point Links

Radio links feeding land mobile systems should follow similar principles to those outlined above for mobile radio systems.

2.12 Effective radiated power

It is in the general interest of all concerned to limit the power output of transmitters to an acceptable level based on recognised criteria. Excessive power is not only usually wasteful but is a recipe for unwanted interference attributable either directly to the transmitters or indirectly to excessive levels in nearby receivers.

Maximum power currently specified by different countries range from a few watts up to kilowatts and whilst the latter may be justified in remote locations, an average level of about 25 watts is generally considered to be the permissible maximum. Even lower powers are recommended in high density urban areas.

2.13 Effective antenna height

This can be defined as the total height of the antenna above average ground level. It therefore includes such height as added by a hill, mountain or building on which the antenna is mounted.

Effective antenna height mainly influences operational range. However, excessive height can be detrimental to frequency planning. The re-use distance of a frequency will be increased as the height is increased. Also spurious products can be radiated or received over a greater distance. Height should therefore be restricted to that necessary to achieve the desired range or coverage.

2.14 Impulsive interference levels

Although not directly a frequency planning parameter, the lack of mandatory noise suppression to vehicles, machinery etc., limits the performance of any land mobile system. The degradation of signal to noise ratio obtainable in urban or industrial areas may dictate higher transmitter powers or additional transmitter/receiver combinations. Reference to the previous points will indicate the effect of such measures.

2.15 Frequency Band

For mobile radio use, it is usual to provide several frequency bands within the VHF and UHF bands. The bands have various characteristics according to their location in the spectrum and it is to the advantage of both users and planners to ensure that the best band is allocated for each purpose.

For extensive rural coverage and, it should be noted, in an electrically quiet environment, the lower VHF frequencies undoubtedly provide best results. For a mixture of urban, suburban and rural districts, the high band (150 MHz) is an ideal compromise. In strictly urban districts, with higher electrical noise levels and greater building concentrations, UHF (450 MHz) is strongly recommended.

2.16 Modulation Method

Land mobile radio systems normally use AM or FM/PM modes. Both modes have their advantages and disadvantages with some of the points resulting in much controversy between different authorities. Both types of modulation have their place in mobile radio. All the above points are discussed in greater detail later in this guide.

3. The Use of the Radio Channel

3.1 Modulation Characteristics

Amplitude Modulation (AM), Frequency Modulation (FM) and Phase Modulation (PM) are the three most well known modulation modes used in land mobile radio services at VHF and UHF (Refs. 3 and 4). The differences between these forms of modulation can be briefly summarised as follows:

(a) *Amplitude Modulation*
A radio frequency carrier wave the amplitude of which is varied in proportion to the instantaneous amplitude of the modulating signal.

(b) *Frequency modulation*
A radio frequency carrier wave the frequency of which is varied in proportion to the instantaneous amplitude of the modulating signal.

(c) *Phase modulation*
A radio frequency carrier wave the phase of which is varied in proportion to the instantaneous amplitude of the modulating signal.

3.2 PM v FM

Of the three modes, AM and PM tend to predominate in the conventional narrow band systems encountered in land mobile radio. The tendency for PM rather than FM in such systems stems from a number of design criteria (Ref. 9), such as:

(a) Phase modulation can be applied directly to a carrier signal and does not require modulation of the carrier oscillator as in FM. Frequency stability is therefore easier to obtain.

(b) Higher deviations can be generated relatively easily with PM.

(c) In PM, the modulation undergoes a 6 dB/octave pre-emphasis in the transmitter and a similar de-emphasis in the receiver (illustrated in Fig. 3.1). The result is an enhanced signal/noise performance, particularly beneficial in weak signal conditions.

3.3 Reducing Channel Bandwidth

In comparing AM and FM/PM we will avoid well-known controversy as to

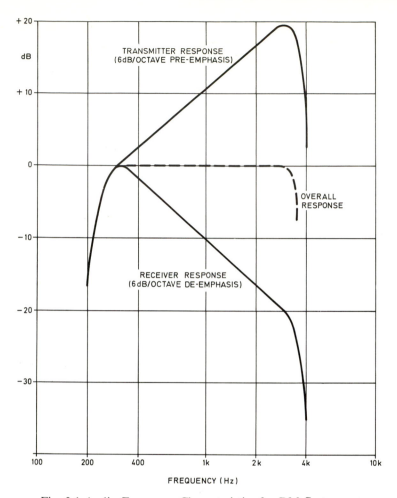

Fig. 3.1 Audio Frequency Characteristics for PM Systems

whether AM or FM is the better system (Ref. 4). It can be seen (Appendix 14) that, with AM, the depth of modulation can remain substantially unaffected when reducing channel bandwidths (there being only a need to prevent over modulation and to limit the top audio frequency cut off). But with FM/PM, the deviation or modulation amplitude must be halved on each occasion the channel bandwidth is halved. Thus, although with wider bandwidths (50 kHz and above) the FM/PM systems show marked system advantages, as the bandwidth is reduced to 25 kHz and still lower to 12½ kHz the advantages rapidly disappear. Tests show that, at 12½ kHz, FM/PM is certainly no better than AM and at limit range AM is markedly superior. If further reductions are likely to, say 6¼ kHz or even 5 kHz, in the future, the rôle of FM/PM may be queried.

Certain disadvantages such as impulsive noise suppression, oscillator stability and filter design are, of course, fundamental to both AM and FM/PM as the channel bandwidth is reduced.

Reducing channel bandwidths in an endeavour to maximise the availability of channels in an expanding land mobile radio field must therefore be considered in detail when planning the spectrum usage in any given set of circumstances. If sufficient spectrum is available to accommodate the needs at, say, 25 kHz bandwidth for the next decade, then it might be considered wise to adopt such a bandwidth. On the other hand the need for subsequent splitting must not be overlooked and therefore the initial channel bandwidth should reflect such a requirement.

In this respect it should be noted that 25 kHz is not the only separation. Certain countries have also allocated channels on the basis of 20 or 30 kHz. Splitting the 20 kHz channel down to 10 kHz implies a more stringent specification than that needed at 12½ kHz channelling. Suitable equipment may be lacking or one may need to accept a degradation in adjacent channel protection. On the other hand, splitting 30 kHz channels results in an eased specification relative to 12½ kHz but gives only 83% as many channels. On this basis, therefore, there is every advantage in maintaining the 100/50/25/12½ kHz sequence.

3.4 Mixing of systems containing different bandwidths

Notwithstanding the conditions we have outlined for orderly channel arrangement, different systems are likely to be mixed initially. Appendix 14 discusses the problems which arise. Appendix 15 considers the stability of the frequency generating sources. It also gives a complete résumé of the quartz crystal, the device which is currently the main means of stable frequency control.

3.5 Modulation Intelligence forms

Within the bandwidths described and used in many countries throughout the world, the actual modulation intelligence takes a number of varied forms.

(*a*)   Mobile radio systems are used predominantly for speech and therefore the radiated bandwidth must be adequate for this purpose.

Speech intelligibility relies upon the transmission and reception of a frequency band containing the bulk of the audio signal energy. It has been shown both experimentally and practically that, provided speech components between 300 and 2500 Hz are reproduced without excessive distortion, intelligibility is adequate for communication.

It must be emphasised that the fidelity of speech using this restricted band of frequencies is not intended to approach that from a broadcasting studio. Indeed, such fidelity would not only require greater bandwidths but also very much higher powers to restore the signal to noise ratios in receivers because of the higher audio frequencies. These higher frequencies contain much of the noise which tends to mar a weak broadcast signal.

(*b*)   A further step in the transmission of intelligence is the use of tone signalling in addition to speech. Selective calling and tone squelch are typical examples of the tone method.

Tones are restricted to three basic types:

(i)   Tone combinations within the speech band and used either before or after speech or in some cases in place of speech.

(ii)   Tone combinations transmitted at any time – either with or without speech – and limited to filtered areas of the audio band, either below 300 Hz or above 2500 Hz.

(iii) Tones inserted in discrete filtered slots in the speech band enabling both speech and tone to be passed at the same time.

These methods should all be possible with conventional designs of mobile radio occupying no greater RF channel width than required for speech-only systems.

(*c*) Later types of transmission, such as tone coding at higher speeds, phase or frequency shift keying, have now become available for data purposes. In many of the systems, the channel bandwidth has remained as for speech although there are some restrictions imposed by so doing.

Speech is often retained even in the more complex data methods as an alternative method of transmission for certain circumstances.

Much has been written on data transmission techniques and so it is not proposed to discuss them in this guide.

The majority of mobile radio transmissions fall into the broad category of speech. So, the channel bandwidths can be clearly defined and wider bandwidths for special purposes confined to a discrete and minimal section of the available spectrum, economising on the available frequency space.

4. Single Frequency Versus Two Frequency Channel Allocations

4.1 General considerations

This is a complex subject. The main points will be discussed here with reference to specialist literature.

In single frequency operation, all transmitters and receivers in a system use the same frequency. Thus all units within range will hear one another.

Two frequency operation usually involves a separation of 4 to 5 MHz with the base transmitter (and mobile receivers) using one frequency and the base receiver (and mobile transmitters) the other. Thus base stations cannot hear other base stations; neither can the mobiles hear other mobiles even within range unless special repeating methods are employed.

4.2 Single frequency operation

At first sight, the single frequency mode seems more economical in frequency spectrum usage. However, a number of disadvantages occur, particularly in areas with considerable concentrations of mobile systems.

As mentioned earlier, (Section 2.3) users who have employed single frequency operation on the VHF bands, have, without exception, changed to two frequency operation in the UHF bands when these bands came into operation.

Let us first of all look at the main advantage in single frequency simplex where the ability of all units within range to hear all other units is of great benefit in sparsely populated areas and in moving convoys. With this method, messages can be relayed if the mobile is in a poor location relative to the main station but close to another more favourably situated unit.

It is for this reason that aeronautical, marine and other safety systems operate in the single frequency mode although its use in the crowded environment of the airfield, for instance, introduces quite restrictive complications.

Now the main disadvantage. If we consider a transmitter in close proximity to a receiver on the same frequency, the level of signal received will be extremely high and can cause adverse effects in the receiver. The high signal level will react quite

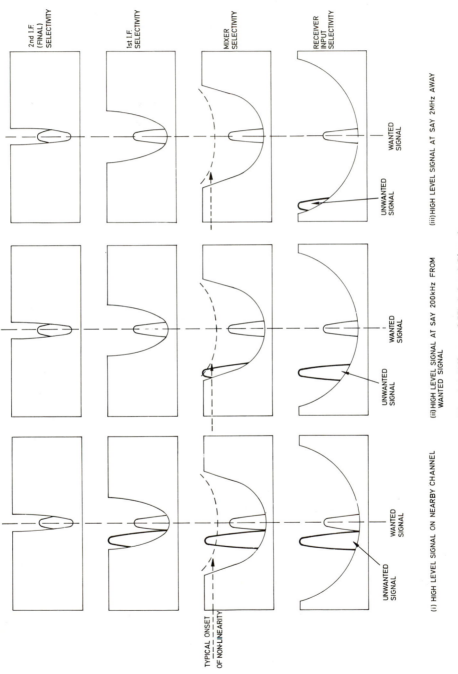

Fig. 4.1 Effects of High-level Signals

violently upon the early stages of the receiver causing non-linearity as well as perhaps temporarily paralysing the first stage. Recovery might well take several seconds and the first part of a message from a weak signal could be missed.

Bearing in mind that output powers of 25 watts are quite common, it can be seen that a receiver input of 5–10 volts could easily be produced by a transmitter in a vehicle alongside.

If amplified by the early stages of the receiver this signal would result in even greater non-linearity by the time it reached the mixer stage.

Identical problems could arise even if the high level signal were on an adjacent channel since the main selectivity is provided by later (I.F.) stages. Fig. 4.1 (i) to (iii) shows how such effects can persist until the early stage selectivity begins to take effect (Fig. 4.1 (iii)). Thus it can be seen that, at a base station, for example, co-sited systems must be separated by an adequate frequency margin (maybe several megahertz) before the effect is reduced to an acceptable level. Thus several such systems on a single site would need this order of frequency spacing *between every system* for adequate protection (see Fig. 4.2).

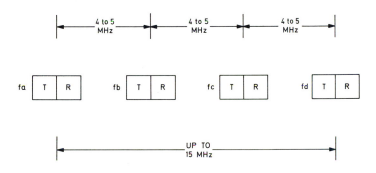

Fig. 4.2 Single Frequency Case – Four Transmitters and Receivers on One Site

This obviously is not feasible in an area containing a concentration of systems in a limited spectrum.

How then can the effects be eliminated?

Clearly one method is to reduce the level of signal on nearby channels to an amount where the adverse effects are removed (Fig. 4.3 (i) and (ii)). To achieve this using only receiver front end selectivity requires the extra attenuation associated with physical separation. Thus we now arrive at the recognised solution in situations involving a number of single frequency channels. Here the transmitters are located on a separate site from the receivers with a sufficient spacing (usually from 1 to 2 km), to reduce the transmitted signals into the receivers to a satisfactorily low level.

With complex systems (aeronautical, marine, etc.,) this is the standard solution. For mobile radio, however, the need for two sites, together with all the associated inter-site wiring or radio links can prove both difficult and expensive and therefore the solution is quite different. It is described in Section 4.3.

Some assistance can be obtained from cavity filters in each receiver input. Whilst this permits a closer physical spacing, the method is expensive and is ineffective if

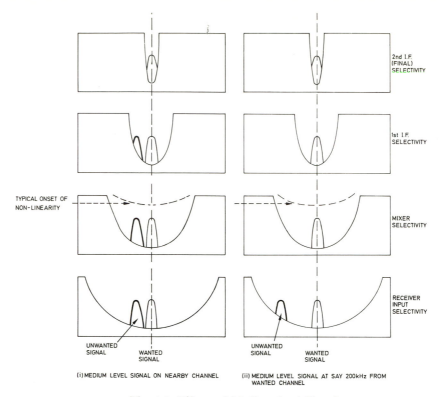

Fig. 4.3 Effects of Medium-level Signals

the physical spacing between transmitters and receivers gives rise to direct pickup of the transmitter by the receiver circuits.

To summarise the single frequency case, the two site method, although the only feasible solution for multichannel operation, is often too complex and expensive for the more conventional mobile radio system.

4.3 Two frequency operation

A development of the ideas described earlier is to group the transmitters and receivers separately giving each group its own block of frequencies. Hence, by separating the two blocks (transmitters and receivers) by a sufficient amount, we have achieved a similar order of protection to that obtained by physical separation in the single frequency mode. The principle is illustrated in Fig. 4.4.

We can see therefore that protection between transmitters and receivers can be obtained by physical separation, frequency separation (or a combination of both).

Some protection can also be achieved by additional selectivity. This usually takes the form of a highly selective duplexer to combine the transmitter and receiver into one antenna whilst providing isolation.

Where conventional transmit to receive frequency spacings are used, such duplexers merely ensure the correct conditions for feeding both the transmitter and receiver into a common antenna. Sufficient selectivity is included to provide similar

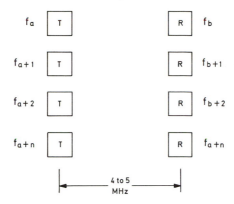

Fig. 4.4 Two Frequency Case – Four Transmitters and Receivers on One Site

conditions to those which would exist if two antennae with normal physical spacings were used.

At reduced frequency spacings however (see Section 5) additional filtering must be included to offset the loss of selectivity in the early stages of the receiver and this is usually added in the form of cavity filters.

Amongst the advantages of two frequency operation is the ability to use the base station in the repeater mode, where, by feeding the audio output of the receiver into the transmitter modulator and arranging for the squelch relay to trigger the transmitter, mobile to mobile communication becomes possible (Fig. 4.5).

In another configuration, the mobile transmit frequency can be continuously monitored by leaving the base receiver operating at all times. Connection into a tele-

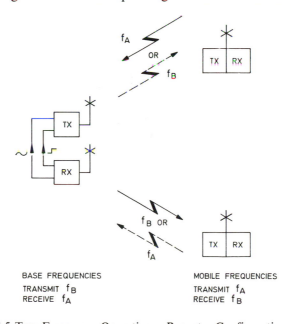

Fig. 4.5 Two Frequency Operation – Repeater Configuration

phone system can be achieved by operating the base transmitter and receiver in the duplex mode.

To summarise, two frequency operation has the advantages of spectrum conservation and easy co-siting of transmitters and receivers. Mobile to mobile communication is not directly possible but can be provided by suitable design, as shown in Section 5. For further details reference should be made to Appendices 12, 13, 19, 25 and 26.

5. The Problems of Mixing Single and Two Frequency Operation in Single and Multiple User Systems

5.1 Introduction

Amongst the various methods advanced as means of conserving spectrum, two frequency operation ranks very high and its use on all possible occasions is considered to be of the utmost importance.

Although two frequency operation is one of the best means of spectrum conservation, it is agreed that certain types of operation may benefit from a limited use of single frequency methods. This section outlines the points to be observed if the maximum use is to be made of the frequencies without inhibiting the use of other channels – whether single and two frequency – in the vicinity.

As we have seen, multiple single frequency systems all require physical separation between transmitters and receivers to operate without mutual interference. Conversely, two frequency systems use frequency spacing to achieve isolation. Hence multiple systems can operate satisfactorily on a single site if the frequency planning has been undertaken correctly.

Single site operation can obviously be used with any types of system, single frequency or two frequency – if only one equipment is located at that site. However, there are occasions when a combination of both single and two frequency operation may be required and the following paragraphs outline the necessary procedures and precautions.

5.2 The Simple Case

On isolated sites, a simple system may need to fulfil a number of mixed requirements, such as:

(*a*) *Base Station*
Capable of being connected into a telephone system, i.e. duplex operation; and/or
Capable of being used as a talkthrough mobile to mobile repeater.

(b)   *Mobiles*
      Operating on a simplex basis to the base station (two frequency operation).
      Operating on a simplex basis to other mobiles through the base station (as
      repeater). Also two frequency operation.
      Operating, when out of base station range, to other nearby mobiles on a
      single frequency simplex basis.

This requirement for both two frequency and single frequency operation in a single
system causes a conflicting state in the mobile equipment. It arises unless one of two
alternative methods is adopted to overcome the switching bandwidth limitation of a
standard unit.

Method A
      In this arrangement, a second RF head is built into the receiver. Thus one head
operates in the allocated mobile receiver block (base transmitter block) of a two
frequency system whilst the second head is capable of operating within the single
frequency block.
      This method relies upon the single frequency block being located adjacent to
the base station receive block and within the mobile transmitter switching bandwidth. This will enable the mobile transmitter to be switched to a suitable single
frequency channel within its nominal frequency switching range.
      Fig. 5.1 illustrates the general frequency plan.

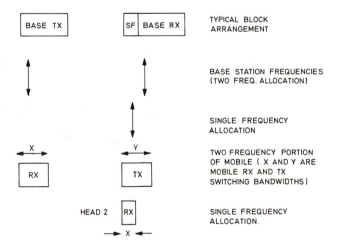

Fig. 5.1 Frequency Plan – Method A

Method B
      Here a special frequency allocation pattern is adopted. Three adjacent blocks are
allocated with the two outer blocks being used for base transmitters and base
receivers. The centre block is reserved for single frequency operation.
      The spacing between the pair of two frequency blocks should be of the order of
500 kHz allowing a *single base station on a single site* to operate in the duplex
mode *provided sufficient antenna filtering is included.*

The blocks should be of a width to permit the mobile transmitter to switch over its own block plus the single frequency block, whilst the mobile receiver should switch over both its receive block and the single frequency block.

Thus the standard mobile can operate on a two frequency basis to its own base station, or on a single frequency basis to other mobiles.

Fig. 5.2 illustrates the general principles.

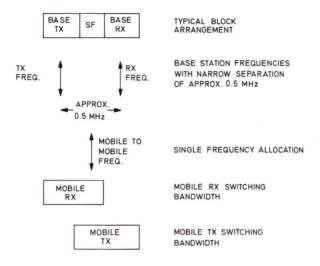

Fig. 5.2 Frequency Plan – Method B

5.3 Multiple Case

Let us first of all consider that the ideal situation exists at a single communal site. This is where all channel allocations are based on two frequency operation with the normal order of frequency separation, 4·5 MHz approximately. With this standard arrangement, all communications from all mobiles to the base stations operate in a two frequency mode.

However, there is no objection to single frequency mobile to mobile communication provided it is a mobile facility and confined strictly to the mobile equipment itself. Method A, where a second RF head is used, is obviously the solution to this problem.

If one now investigates how to use *standard* mobiles in the method B configuration, then immediately a base station problem must be recognised. In the simple case (5.2 above), duplexers for this purpose would fall into the band-reject category (see Appendix 11) and will necessarily exhibit selectivity characteristics which will change rapidly as an incoming signal moves off tune. Therefore, although the performance at the tuned frequency can be excellent, within a channel or two of the tuned point the efficiency of the duplexer starts to drop rapidly.

This means that for one system, method B can be used effectively. If, however, a second system in the same band is installed on the same site, then the ability of the duplexer of System 1 to reject the transmitter frequency of System 2 and vice versa may be completely nullified and interference will result. The presence of other equipments on the site will merely worsen the situation. So we no longer have isolation.

Whilst special filtering can sometimes be employed, this may, even if possible, be extremely expensive.

To achieve the required level of base station transmitter to receiver isolation we must revert to the same method necessary with single frequency operation. That is to provide physical separation between transmitters and receivers, the distance being a function of the degree of isolation required, bearing in mind blocking and intermodulation possibilities.

It is important to realise that the transmitter to receiver physical spacing to minimise intermodulation must necessarily be greater than for blocking. With close frequency separation between receivers and transmitters the probability of low order on-channel transmitter products falling on receive channels is relatively high. More will be given on this subject in later sections. Fig. 5.3 briefly shows the problem but does not take account of the difficulties of adequate filtering when two or more systems of this type are in close proximity.

By separating the sites, therefore, we have achieved one important advantage, mobile equipment simplicity. By using close frequency spacing, standard single head type mobiles can be used – the second RF head being avoided.

However, an exception arises with some established systems, the International Marine System being a typical example. Close frequency spacing is not used in this service and the two frequency mode is used at 4·6 MHz spacing. However, single frequency operation is also needed by the *Base Station*. So not only is a double headed mobile required for the system, but also the base stations associated with the worldwide system must include separated transmitter and receiver sites to ensure that interference does not affect the land-based receivers in the various operational modes practised.

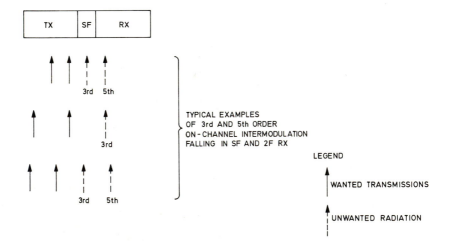

Fig. 5.3 Intermodulation Problem with Multi-system Close-spaced Two
Frequency/Single Frequency System

FREQUENCY ENGINEERING IN MOBILE RADIO BANDS

William M Pannell C.Eng., M.I.E.R.E

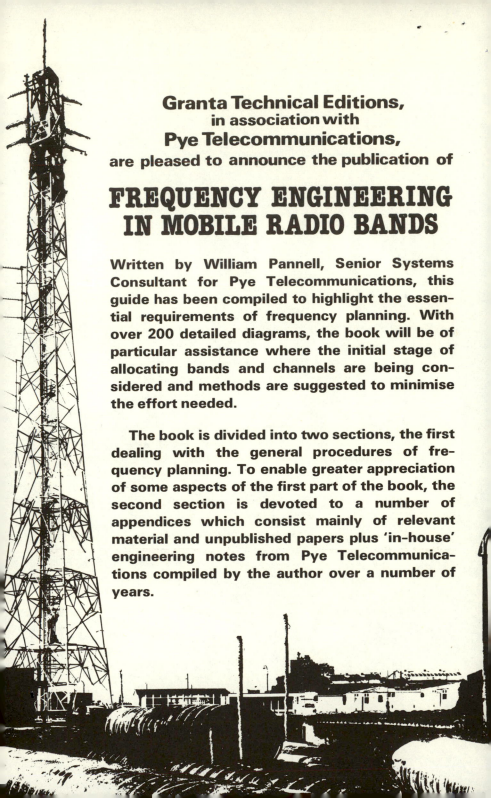

Granta Technical Editions,
in association with
Pye Telecommunications,
are pleased to announce the publication of

FREQUENCY ENGINEERING IN MOBILE RADIO BANDS

Written by William Pannell, Senior Systems Consultant for Pye Telecommunications, this guide has been compiled to highlight the essential requirements of frequency planning. With over 200 detailed diagrams, the book will be of particular assistance where the initial stage of allocating bands and channels are being considered and methods are suggested to minimise the effort needed.

The book is divided into two sections, the first dealing with the general procedures of frequency planning. To enable greater appreciation of some aspects of the first part of the book, the second section is devoted to a number of appendices which consist mainly of relevant material and unpublished papers plus 'in-house' engineering notes from Pye Telecommunications compiled by the author over a number of years.

William M Pannell C. Eng., M. I. E. R. E

William Pannell has 47 years of electronic experience, most of the time in Mobile Radio. Educated in the City of Cambridge, he joined Pye Telecommunications Ltd (then Pye Radio) in 1932, working in the Research Laboratories.

During the years of World War II he took over the section which designed the two well known Army wireless sets, WS22 and WS62, the latter being used in the first crossing of the Rhine by the British Airborne Division.

After the war he began working on domestic radio development followed by mobile communications equipment and in 1957 started the Systems Engineering Department of Pye Telecommunications Ltd. From 1965 he was the technical manager for Overseas Marketing and was closely involved in projects which included a VHF multiplex link system for aeronautical use in all major islands of the Caribbean; a complete airport and marine system in Basrah Ports in Iraq; a security system for the United Arab Republic; and a security system in Rio de Janeiro, Brazil.

In 1972 he took up his present appointment of full-time Systems Consultant to the Company. The author of some 30 published papers, Bill Pannell is responsible for 'The Pannell Report - The Future Frequency Spectrum Requirements for Private Mobile Radio in the United Kingdom'. The author is married and lives in Stapleford, near Cambridge. He has two sons, one in electronics and the other in instrument engineering in the USA.

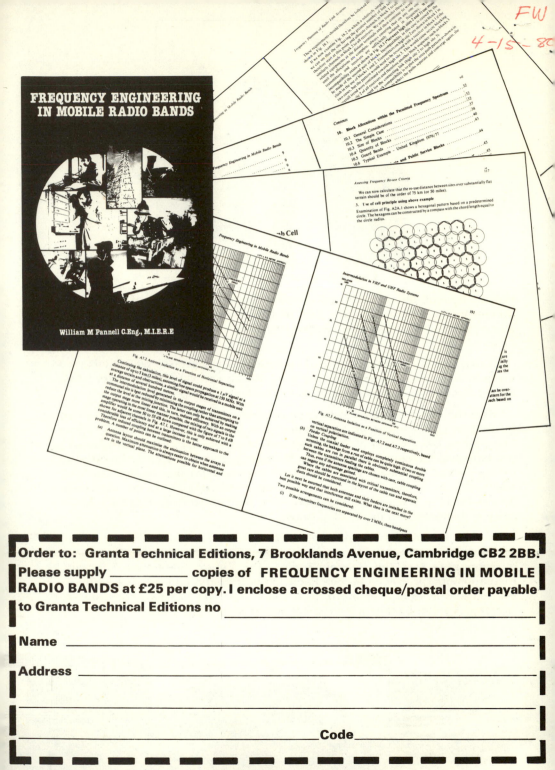

6. Types of System Normally Encountered

Although at first, the term "private mobile radio" suggests the conventional type of communication system involving one or more base stations and the associated mobiles or portables, we must not lose sight of the other types of application when frequency planning principles are being determined.

Let us therefore review the various applications which are broadly contained within the description "private mobile radio".

6.1 Wide Area Communication Systems

These form the backbone of many private mobile radio systems and consist mainly of base stations with associated mobiles or portables. Transmitter powers and antenna heights are such as to provide coverage over a wide area and frequency re-use is consequently limited. Frequencies in the VHF bands are normally used.

6.2 Local coverage systems

These often fall into two categories, one fulfilling a function such as coverage within a city or part of a city and the other, of a more limited nature, of covering a small town, factory site or similar area having rather more defined boundaries.

Both types however follow similar principles, the main differences being that of permitted power and antenna heights. These tend to be limited and therefore frequency re-use can be practised on a reasonable scale. Frequencies in the UHF bands are usual for this type of system.

6.3 Radiating cable systems

This form of communication is used in areas where conventional forms of radiation cannot be used. Appendix 16 describes the engineering considerations. For example, in subways, lift and mine shafts, underground shopping precincts and railway tunnels, the range covered by conventional antenna systems is severely reduced and would only extend a matter of metres from the antenna. However, by running a length of coaxial feeder along the route to be covered and allowing this feeder to leak or, more correctly, radiate along its length, communication can be achieved with mobiles or portables in the vicinity of the cable. Similarly the cable will pick up signals along its length and pass them back to a receiver to enable a talk back path to be established.

By virtue of its limited range, very few frequencies are needed as they can be re-used within very short distances. Both VHF and UHF can be employed but where extended coverage is required, VHF is preferred as the operational cable length is a function of its loss. The loss varies as the square root of the frequency (e.g. cable power loss in dB at 400 MHz is twice that at 100 MHz for the same grade of cable).

Where radiating cable systems marry into conventional systems – for example a system involving tunnels along a normal railway route, or a tunnel in a conventional road system – the frequency plans must be compatible. Since radiating cable systems are unlikely to cause any interference to conventional systems or vice versa, frequencies from the bands reserved for normal systems can be allocated without difficulty to cable systems. Even using the same channel, the transition distortion at the junction of an open and a subway section is unlikely to extend more than a few metres.

Cables can also be used to carry a number of frequencies simultaneously, provided suitable combining and filtering arrangements are included in the overall network.

6.4 Radio Links

In this category we are mainly concerned with the simple point to point link involving a single speech circuit. However, other types of link, over which a number of speech circuits are multiplexed, must be mentioned as they often form a considerable part of a country's communications network. High capacity radio links involving hundreds of speech circuits are always accommodated in bands above 2 GHz, but lower capacity links (carrying up to 60 channels) are often allocated frequencies in the 450, 960, 1500, or 1900 MHz bands. The trend, however, is to allocate in the upper ranges, particularly if 450 MHz is extensively used for private mobile radio. In the UK, single channel point to point link equipment, for some years allocated channels in the 450 MHz band, is now being designed for use in the 1500 MHz band. The intention is to use the vacated sections of the 450 MHz band to extend the mobile radio capability.

Whichever band is used for link purposes, however, with directional antennae, re-use can be noticeably improved. Also, using a different band from mobile radio, the interference potential is greatly reduced.

Section 18 discusses frequency planning for radio links.

6.5 Paging

Paging methods fall into a number of categories. In this guide we are mainly concerned with those occupying the VHF and UHF ranges. Those using inductive loops and the lower frequency bands are, therefore, not considered.

Paging systems in the VHF and UHF bands can be sub-divided into:

  (*a*)   Restricted area low capacity systems;
  (*b*)   Medium area, medium capacity systems;
  (*c*)   National, high capacity systems.

The first category usually employ frequencies in the UHF bands and cover limited areas such as factory sites and hospitals. The internal telephone system is used for replies.

The second category involves larger areas such as towns. In one configuration – overlay paging – this category forms part of a conventional radio speech system. Here the pager can be used as a form of personal selective call unit with the recipient using his vehicle radio (as well as the local telephone system) for answer back purposes. VHF is invariably used in this arrangement.

The national system, available in certain countries, allows its users to be called whilst they are in any part of the country. The answer back is by telephone. Transmitters are located at strategic points throughout the areas to be covered.

Some paging systems, particularly those having limited areas of coverage, enable the alerted unit to be given a short message. This type of system tends to be limited to hospitals etc., where the called person is required urgently at a location which can be announced immediately after the alert tone.

Other systems provide a radio talk back path but this type of system is better categorised as a conventional local area two way communication system using selective calling facilities.

Paging systems often use radiating feeder techniques especially where coverage is required in buildings and where screening prohibits the use of conventional antenna systems. Mixtures of radiating cables and conventional antennae are also often used in some system configurations.

7. Preferred Frequency Bands

The choice of frequency bands for private mobile radio depends upon a number of factors amongst which are:

(*a*)   Size of country
(*b*)   Type of terrain
(*c*)   Vegetation cover
(*d*)   Type of climate
(*e*)   Degree of urbanisation
(*f*)   Noise environment

7.1   Size of Country

The size of the country tends to dictate the frequency bands *not* to use rather than the reverse. For example, in a small country, surrounded by other countries, the longer ranges which can be achieved by using, say, frequencies in the 40 MHz band, would not be in the best interests of all the countries concerned. On the other hand, a country with vast tracts of desert may not need coverage of those areas and therefore again the lower frequencies may not be necessary. The longer ranges of the lower frequencies tend to be best suited to large countries with distributed populations rather than isolated population areas in an otherwise uninhabited wilderness. The United States, Canada, parts of Africa and Australia are the best examples using the lower end of the VHF band.

7.2   Terrain

The effect of terrain on the choice of frequencies is somewhat complex. Rolling terrain, for example, tends to be better served by lower frequencies, whilst with some types of mountainous area, it is often advantageous to use higher frequencies. Coverage of mountainous areas however, often requires special treatment. For example, valleys can usually be better covered by antenna systems in the higher frequency bands firing along a valley rather than an antenna located on a high site trying to illuminate a large area, often shadowed by other peaks.

7.3 Vegetation

Attenuation due to vegetation coverage is extremely frequency dependent. Light scrub or open cultivated areas have very little effect upon the propagation losses in the normal mobile radio bands. As the vegetation thickens, however, ultimately becoming thick forest, the attenuation becomes quite high, particularly at the higher frequencies. For example, an antenna, below and surrounded by trees, might show a loss of 2 to 3 dB at 30 MHz relative to an open location. This loss could worsen to 5 to 10 dB at 100 MHz, whilst, above this frequency, losses could increase even more rapidly. At 1000 MHz, a visually opaque mass would exhibit an extremely high loss.

7.4 Climate

If we consider the effect of climate on vegetation, then the effect will change considerably with seasons. In general the direct losses will increase as the obstructions become more saturated with water or the foliage becomes thicker.

However, obstructions such as mountains or buildings, if saturated with moisture, often provide effective reflecting surfaces. Whilst they affect the direct radio path adversely, the reflecting path is often improved. So the overall coverage is modified to a marked degree.

Fog can increase the losses in the microwave region but has very little direct effect at mobile radio frequencies. However, fog often results in a temperature inversion causing temporarily enhanced coverage and interference can be caused over quite long distances, particularly at VHF.

7.5 Building effects

The choice of the frequency band for a built up area is complicated issue. Amongst the factors to be considered are the size and shape of the area, types of building, the degree of electrical noise in the area, and the types of system needed.

In areas containing many modern buildings, notably reinforced concrete structures, the use of 450 MHz is recommended. This frequency band is not only capable of enhanced reflection from such structures but also penetrates well into buildings and areas normally exhibiting high attenuation at the lower frequencies. Furthermore, even in a high electrical noise area, this particular band is much less affected. C.C.I.R. show that the city noise at 450 MHz is often less than rural noise at 100 MHz. Figure 7.1 illustrates this data, taken from C.C.I.R. Study Group Document DOC 6/167E/F/S. However, there may well be considerable disparity in noise levels between similar areas in different countries owing to variations in suppression conditions and laws.

7.6 General considerations

The 450 MHz band is capable of covering urban areas well although in suburban or rural areas there is a marked loss of coverage compared with the lower frequencies. The causes are mainly vegetation loss, fewer and poorer reflection points and the greater free space loss expected at higher frequencies.

Multiple high sites tend to help matters when urban, suburban and rural coverage are all needed but often the compromise of using frequencies in the 150 MHz band is to be preferred.

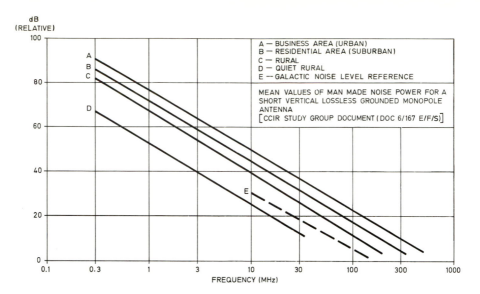

Fig. 7.1 Mean values of Man-made Noise Power for a Short Vertical Lossless Grounded Mobile Antenna

The types of system needed often dictate the band of frequencies to use. Limited coverage systems using pocket equipments imply the UHF band as the most suitable. An area system, extending over several thousand square kilometres using many types of vehicle, is best served by VHF, preferably below 100 MHz.

Appendix 24 gives further information on the subject.

8. Sharing of Mobile Radio Bands With Other Services

The sharing of particular bands by two or more services is often practised in countries where there is sufficient protection between the services. This protection is invariably provided by geographical features such as mountains, tracts of desert, forest, sea or simply distance.

Typical amongst the sharing possibilities are:

(a) Parts of the marine band may be employed for purposes such as land mobile in locations remote from the nearest coastline or navigable river.

(b) Parts of the TV or broadcast bands can be used for other purposes in areas in which TV or broadcast reception is not wanted or possible and unlikely to be required in the foreseeable future.

Agreement must be obtained from the relevant authorities before sharing is put into operation. Consultations with neighbouring countries may be necessary and any interference arising from sharing must, of course, be cleared by the secondary service users.

Sharing of services must be approached with care with limits imposed on transmitter powers and antenna heights if interference is likely. The possibility of anomalous propagation must be borne in mind when deciding the system's parameters.

Higher power, although effective mainly within the area of coverage, does increase the range slightly by offsetting path attenuation. However, within the main coverage area, excessive power can cause interference. It can cause large unwanted signals in other nearby receivers and it can, in conjunction with other high power transmitters, produce intermodulation products on wanted channels. Furthermore, the levels of unwanted harmonic radiation can be unduly high.

Thus any transmitter power output should be adequate for satisfactory signal levels at the limit of required coverage. It should not be increased to offset, say, the poor location of a mobile receiver near the perimeter of the coverage.

Power levels on communal sites have to be carefully considered in view of the requirements and the possibility of interference.

Increasing transmitter power to combat high level electrical noise is also deprecated. This is one of the most convincing arguments for mandatory electrical noise suppression. Merely raising the transmitter output to overcome high levels of man made noise is self-defeating, as the ultimate result is to create still more pollution.

9. The Effect of Antenna Height and Transmitter Power

Ever since the inception of radio communication, maximum antenna height and high transmitter powers have been advanced as essential to the best service. Private mobile radio is no exception to the trend but it must be realised, however, that spectrum usage is not only a function of the amount of frequency occupied but also of the area over which that amount of frequency is irretrievably "lost" to other users.

Three basic "dimensions" exist in respect of frequency usage and these can be broadly classified as frequency, amplitude and time. Appendix 4 describes this approach in greater detail and it is therefore sufficient at this stage to point out that a channel, comprising a carrier and its sidebands, should only cover the operational area of the required system and not extend to any marked degree beyond this limit. Nor should its occupancy in the time dimension be excessive in view of other users in the same area who may need the same channel during the unused periods.

Reverting to the "dimension" of distance, it is clear that maximum usage of a channel is consistent with the greatest re-use of that channel within an overall area. In turn, re-use depends mainly upon the antenna height and partly upon the transmitter power. Thus it is in each user's interest to ensure that their antenna height is sufficient but that excessive coverage is avoided.

10.   Block Allocations Within the Permitted Frequency Spectrum

10.1   General considerations

As discussed in Section 4 the spectrum must be sub-divided for the greatest frequency economy with the minimum interference.

It has been shown that two frequency operation provides the basis for optimum utilisation of the spectrum and, although single frequency operation is not ruled out for certain applications, its use should be strictly controlled with full observance of its interference-creating potential.

For the two frequency case, blocks of channels, adequately separated, should be allocated in a pattern chosen for maximum transmitter to receiver isolation. The pattern will obviously depend upon the available frequency spectrum, its arrangement and the national pattern of users' needs.

In the final scheme, sufficient guard bands should exist, not only between the various blocks but also as a protection against services in adjacent segments. Single frequency segments should only be included with due regard to all users.

Care must also be taken to avoid sharing the bands with special types of radio service which, although allied to private mobile radio, may have widely differing characteristics. For example, wide band transmissions with multiplexed speech channels must not be included. They may even be part of the same system, e.g. feeding several speech circuits to a communal site. But their inclusion is fundamentally unsound, leading to wasteful spectrum usage or unacceptable interference. Channels for links or wide band transmissions should be allocated within bands exclusively reserved for such services – usually in the higher bands above 900 MHz. The sharing of the mobile radio band with non-allied transmissions, such as programme links or music links for broadcast purposes, should be similarly banned.

10.2   The simple case

Let us first of all look at a simple case satisfying the main rules of two frequency operation. Intermodulation will not be considered in detail in this simple case.

Observing strict rules to avoid on-channel intermodulation is necessary when allocating individual channels but, when allocating blocks, only the general case needs to be considered.

Fig. 10.1 shows two blocks of frequencies, one for transmit purposes and the other for receiver use. Adequate spacing must exist between the two blocks to provide sufficient isolation. In particular, this means the separations between Channel 1 Tx and Channel 1 Rx, Channel 2 Tx and Channel 2 Rx and so on, must exceed, for example, 4·5 MHz. Tx to Rx differences of I.F. or half I.F. (usually 10·7 or 5·35 MHz) should be avoided. The blocks should, of course, be planned in relation to other blocks, other users and other services.

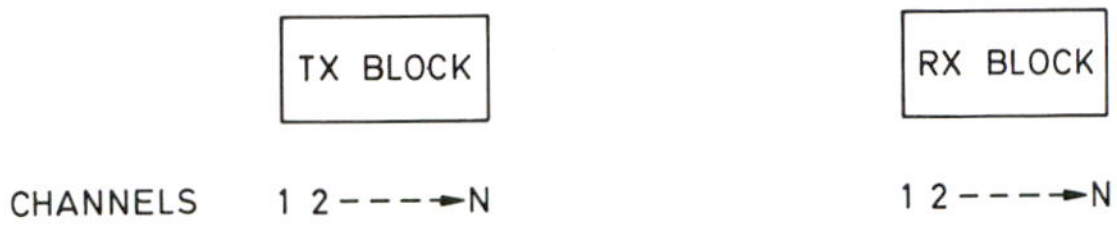

TX BLOCK	RX BLOCK

CHANNELS   1 2 – – – –►N   1 2 – – – –►N

Fig. 10.1 Main Requirements for Two-frequency Block Arrangement

Let us analyse the problem by stages.

(a) *Blocking or desensitisation*
     This problem depends upon the received signal level and is a function of the selectivity of the early stage of the receiver. In effect, the receiver sensitivity is reduced.

(b) *Isolation requirements*
     For signals close to the tuned frequency of the receiver, blocking affects the receiver sensitivity to wanted low level signals when the interference signal input approaches 20 mV.

     A 25 watt transmitter generates 35 volts across a 50 ohm load (the antenna) and therefore, if this voltage is not to cause blocking in a receiver tuned to a *nearby* channel within the front end selectivity response, there must exist an attenuation of

     $$20 \log_{10} \frac{35}{0 \cdot 02} \ \text{dB}$$

between the transmitter output and the receiver input. This is approximately 65 dB. Figure 10.2 shows the isolation obtainable from vertical separation of antennae.

     If we assume a space attenuation of 30 dB on a single site between transmitter and receiver antennae, then we require 35 dB additional attenuation to prevent receiver blocking.

     Examination of Figure 10.3 (the front end selectivity of a conventional receiver) shows that 35 dB can be achieved if the transmitter is spaced by 3% from the receiver tuned frequency. At 150 MHz this implies a spacing of at least 4·5 MHz.

(c) *Effect of transmitter noise on block spacing*
     A radiating transmitter, whether modulated by intelligence or not, also carries a spectrum of noise which is mainly caused by random fluctuations

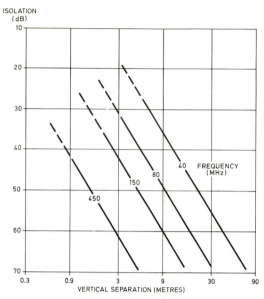

Fig. 10.2 Antenna Isolation as a Function of Vertical Separation

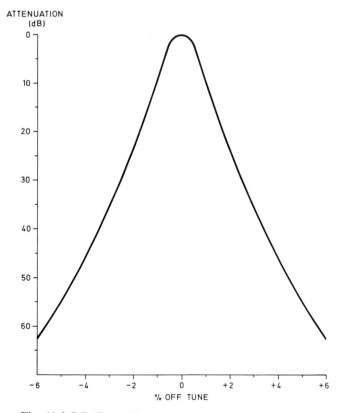

Fig. 10.3 RF (Front End) Selectivity of a Typical Receiver

of current flow, usually in the frequency generating sources of the unit. This noise spectrum, which spreads for some megahertz on either side of the carrier, is another reason for adequate isolation between transmitters and receivers.

Measurements indicate that transmitted noise is about 45 dB below carrier level at the transmitter frequency, falling to 80 dB at 15 kHz from the carrier and thereafter dropping by the rate of 4 to 6 dB per octave. As will be seen in Appendix 25, this indicates 140 dB isolation at frequencies around the spacing required for the elimination of blocking.

It must be emphasised that transmitter noise is completely unaffected by receiver selectivity or by any additional filters fitted to the input of the receiver. It can only be further reduced in level by filtering the *transmitter* output – either using a bandpass filter or a notch filter depending on the frequency separation (Tx – Rx) and the number of receivers to be protected. Obviously a notch filter will protect only a receiver to which the notch is tuned, whilst a bandpass filter will protect a number of receivers provided the frequency spacing allows adequate attenuation by the filter. Appendix 11 discusses the isolation requirements affecting duplexers. Similar techniques using single or multiple coaxial filters must be employed when separate antennae are used with reduced frequency spacings. (See Appendix 25).

As shown in the case of blocking, with normal power levels, a separation of 4·5 MHz also provides sufficient protection against transmitter noise without additional filtering. Care must, however, be taken on communal sites, as shown in Fig. 10.4 (the blocks, all 1 MHz wide, all refer to base stations). In the example, a Tx to Rx spacing of 4·5 MHz is assumed

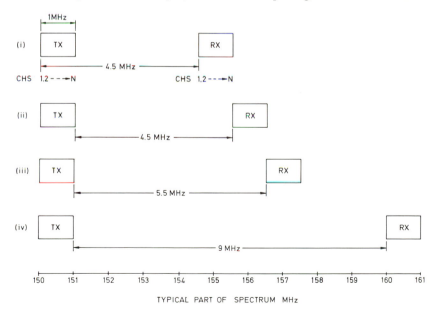

Fig. 10.4 The Effect of Different Transmitter to Receiver Block Spacings

necessary for each channel. But, in case (i), Channel N Tx to Channel 1 Rx would be only 3·5 MHz away (possibly on the same site). Also, half IF separation could also arise (see Appendix 26). The first problem is avoided in case (ii), but the second is still a possibility. The further separation of case (iii) eliminates both problems. Case (iv) is another suitable plan, although it may introduce 10·7 MHz differences.

Fig. 10.5 shows how this effect can be worsened still further if the block size is excessive and emphasises the need for careful planning of both block size and spacing, particularly if communal sites are to be the rule rather than the exception. In Fig. 10.5, all blocks are 3·5 MHz wide. In case (i), although basic requirements are fulfilled, problems can arise on communal sites, including Tx to Rx spacings as low as 1 MHz (Channel N to Channel 1), IF-related Tx to Rx differences, third- and higher-order inter-modulation products in the Rx block and excessive Tx noise. In case (ii), low spacing is avoided, but IF-related differences and high-order inter-modulation products (5th) may be troublesome without careful planning. Clearly, therefore, the block size is excessive and sub-division is to be preferred.

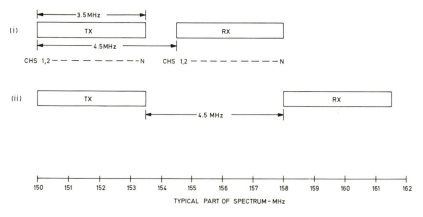

Fig. 10.5 The Effect of Excessive (Base Station) Block Size

(d) *Comparison with single frequency operation*
By adequate separation of the two blocks we have therefore not only achieved sufficient isolation to prevent desensitisation of nearby receivers listening to weak wanted signals, but have also ensured that incoming signals are not marred by transmitter noise. In a single frequency simplex case, the required total isolation of 65 dB to avoid blocking would have needed a transmitter to receiver spacing of 500 metres when operating on nearby channels (see Fig. 10.6). The physical spacing required for sufficient protection against transmitted noise for a receiver on a channel adjacent to the transmitter generating the noise would need to be roughly twice this distance.

It could be argued by examining Fig. 10.2 that the whole of the required attenuation could have been achieved by vertical separation between the transmitter and receiver antennae. This is theoretically possible but

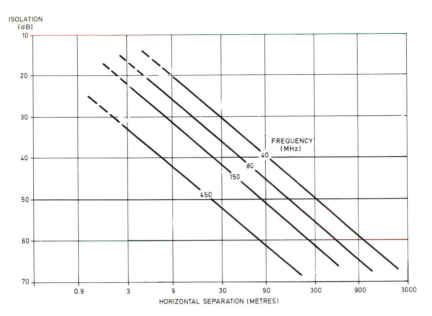

Fig. 10.6 Antenna Isolation as a Function of Horizontal Separation

difficult at the higher levels of required attenuation, owing to feeder coupling, direct unit to unit hop-over etc. Additionally a separation of over 40 feet between *every* antenna on a mast, plus the need for all antennae to be truly collinear makes a practical solution extremely unlikely, even if one only considers mast size as the limiting factor.

10.3   Size of blocks

Several factors dictate the size of each block.

(a)   The individual block must be wide enough to accommodate a reasonable number of channels based on the bandwidth adopted.

(b)   The block must not be excessively large otherwise it is often difficult to locate the blocks in the available spectrum at the required spacing and also to achieve adequate transmitter to receiver spacing with some allocation patterns.

(c)   As seen in Fig. 10.5, the probability of transmitter lower order inter-modulation products falling in the receiver band is much greater as the block size increases.

(d)   Certain transmit to receive frequency differences are best avoided, particularly on communal sites, if interference is to be minimised. It is impossible, universally, to cater for all the pecularities of individual manu-facturers' equipments. But differences directly related to internationally used frequencies, such as first and second IF, are best avoided (see Appendix 26). This is mainly achieved by individual analysis of the requirements and the possible solutions for each site.

Reference to Figs. 10.4 and 10.5 show such effects and various alternative solutions.

10.4   Quantity of Blocks

For the reasons given above it may, for instance, often be beneficial to sub-divide every particular part of the spectrum into more than a single pair of blocks in order to permit more than one combination of channel configuration to be used on crowded sites.

Many such arrangements are possible and depend mainly upon the amount of spectrum available, the neighbouring services, the guard bands needed, and the general distribution of the channels required throughout the country. A number of variations are possible: Figs. 10.7/8/9 indicate a few possibilites. No effort is made to show all the parameters; only those requiring attention being highlighted.

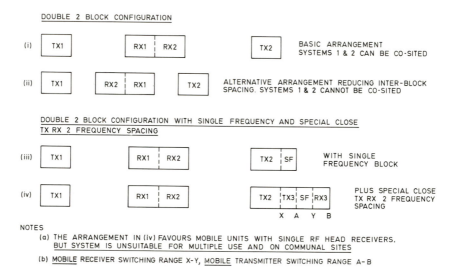

Fig. 10.7 Double 2 Block Configurations

We discussed in Section 5 the occasional need for both single frequency and two frequency operation in a single system and described the two methods used (twin headed or wide band receiver technique and close transmitter and receiver two frequency spacing). Therefore some of the alternative frequency plan arrangements include blocks which provide for these special forms of communication.

Fig. 10.7 shows first a double two-block plan in which both of the base station receiver blocks are located between the two transmitter blocks. The blocks shown in the figure refer to base stations.

The advantages are clear. First, two plans are located in a total spectrum of approximately 1½ times that of a single plan.

Second, by locating the transmitters at the outer blocks, transmitter inter-modulation products resulting from combinations from each of the two blocks will be very much reduced in amplitude and *cannot* fall in the receiver bands. Products generated as a result of two or more transmitters in any one block will be similar to those generated in the simple two block case, falling near to the transmitted channels, with only low amplitude high orders nearer to the receivers.

In the last example of Fig. 10.7, it can be seen that an extension has been made to the second transmitter block (Tx2) and this extension (Tx3) operates together with the Rx 3 block which, as can be seen, is separated from Tx3 by a single frequency block. Thus, with this special arrangement (transmitter to receiver spacing about 500 kHz), two-frequency operation is possible with special filtering at single base stations, whilst the mobile will be able to switch to the single frequency band for mobile to mobile operation. The spacing between the start (X) of Tx3 and the end (Y) of SF must be within the frequency switching range of the mobile receiver. The mobile transmitter frequency switching range must fall between the start (A) of SF and the end of Rx 3. Section 5 refers to this special method of single and two frequency operation.

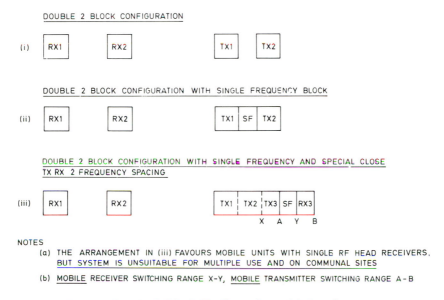

NOTES
(a) THE ARRANGEMENT IN (iii) FAVOURS MOBILE UNITS WITH SINGLE RF HEAD RECEIVERS, BUT SYSTEM IS UNSUITABLE FOR MULTIPLE USE AND ON COMMUNAL SITES

(b) MOBILE RECEIVER SWITCHING RANGE X-Y, MOBILE TRANSMITTER SWITCHING RANGE A-B

Fig. 10.8 Double 2 Block Configurations with Restricted Spectrum

With insufficient spectrum for such a single integrated block system, a solution is possible if several small segments are available. This solution is illustrated in Fig. 10.8, where the blocks shown all refer to base stations.

Here the separation between the transmit and receive blocks fall in with the isolation requirements previously discussed, whilst transmitter intermodulation products in the receiver band are no worse than in the case of the single pair of blocks.

Experience has shown that this configuration often results in greater spacings between transmitter and receiver blocks than in the single pair case. This is because the smaller blocks are often located at random to fill the spaces between services rather than the reverse.

Fig. 10.8 shows, in the second example, the location of a single frequency block, whilst the third configuration shows the addition of a close-spaced two frequency section located on either side of the single frequency sector. This arrangement is similar to that of Fig. 10.7 and is suitable for the special arrangement previously discussed.

Fig. 10.9 shows an alternative position for a single frequency block – a configuration requiring a double RF headed receiver, as in international marine service. A close physical spacing between base transmitters and receivers is not recommended when the single frequency block is near to the receiver block owing to the greater possibility of interference to the multitude of two frequency receivers in a crowded area.

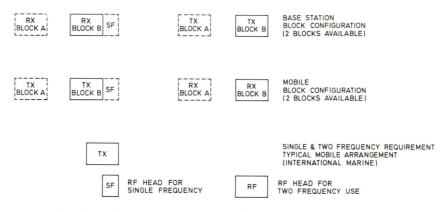

Fig. 10.9 Alternative Location for Single Frequency Block

10.5   Guard Bands

Having arranged the layout of the blocks to provide suitable isolation between transmitters and receivers, three possibilities remain in respect of the spectrum or guard bands between blocks:

(*a*)   If the guard bands are part of the spectrum allocated to *private* mobile radio, then the use for these spaces must be carefully considered.

(*b*)   If they are sections of spectrum allocated to public mobile radio, then similar arrangements to those for private mobile radio – suitable blocks, etc. – must be made, with sufficient physical spacing between the private and public systems to ensure adequate isolation.

(*c*)   If they are allocated to other services, care must be taken in the arrangement of those channels in the mobile radio section which are physically near to transmitters of the other services located in the guard bands.

Fortunately, in the last case, I.T.U. planning generally avoids widely incompatible services being allocated on nearby frequencies. If, however, mobile radio is next to, for example, TV or other wideband services, the number of interference sources is limited with strictly defined locations applying. Thus it is generally not difficult to provide physical isolation in such particular cases.

In the second case, where public blocks are arranged, so far as possible, not only for isolation in their own right, but also to avoid problems in the private sector, we again tend to rely upon physical separations. Public systems are generally installed in special localities and so the likelihood of common siting of private and public systems can, with care, be avoided.

The first case, although at first sight appearing to confuse the recognised block system, can often be used for special systems in isolated areas. For example, in a country with vast sparsely populated tracts, radio sites are most likely to have only one transmitter and receiver each, unlike areas of greater density. Under these conditions, single frequency allocations in the guard bands can often be considered, provided of course that adequate protection is observed near the junctions with areas having a greater density of units.

The special close-spaced two frequency system described in Section 5.3 can operate very well in these areas.

Even in countries having high user densities, the guard band can be occupied by confining its use to areas of the country having naturally isolated regions. For instance, a peninsular, valley between mountain ranges or island(s) may often be sufficiently distant from the main areas to provide the required isolation.

It must not be construed from the above that the recommended block arrangement should be completely ignored within guard bands. Undoubtedly, two or more sections of the spectrum can often be combined to provide a suitable two frequency block arrangement. For instance, although we have indicated block spacings of the order of 4·5 MHz as generally suitable, there is no reason why several tens of megahertz should not exist between the transmit and receiver blocks. The main reservation is the need to use two antennae – one transmit and one receive – at the base station, with the mobile antenna being tuned to the mobile *transmitter* – a resonant antenna for the mobile receiver being of less importance.

Some lack of range reciprocity may however exist, with a slight reduction in range at the higher frequencies. However, if the base transmitter is allocated the higher frequency, together with a slightly higher power than that of the mobile, the difference will tend to be ironed out. Commonly, a 25 watt base transmitter and a 10 to 15 watt mobile transmitter are used, achieving the required differential.

Under certain circumstances it may be considered that a small segment of the band should be left unallocated to divide two services having conflicting interests. Whilst not appearing in the true interests of frequency economy, it could be that spurious effects falling in the occupied bands may be thus avoided with a consequently greater use of the spectrum in the area involved. For some scientific applications, this method may be essential if all users are to benefit. Certain wide band devices such as diathermy and RF heating may necessitate a sterile guard band in busy locations having a high mobile radio density.

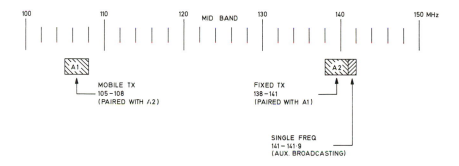

Fig. 10.10   VHF Mid-Band Spectrum Available in UK for Private Mobile Radio

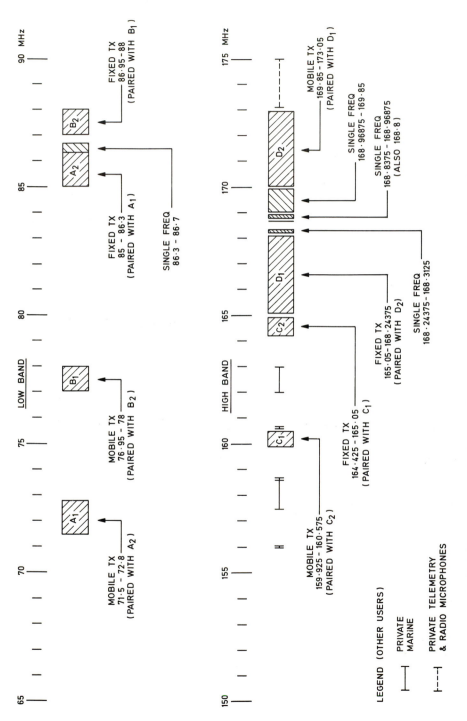

Fig. 10.11 VHF Low- and High-Band Spectrum Available in U.K. for Private Mobile radio

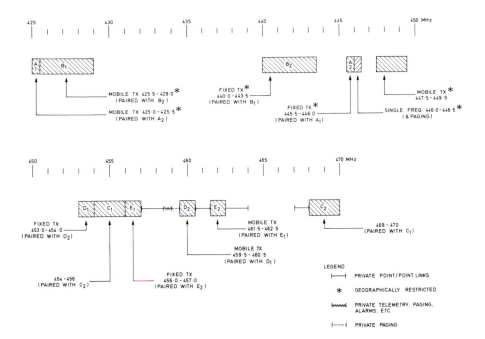

Fig. 10.12 UHF Spectrum Available in U.K. for Private Mobile Radio

All these restrictions, however, necessarily hinge on the protection incorporated in the initial planning and the arrangement of the various services within the overall spectrum.

10.6   Typical example – United Kingdom 1976/77

As an example, consider the current (1976/77) private mobile allocation in the United Kingdom. Figs. 10.10 and 10.11 show the VHF allocations and Fig. 10.12 shows the pattern of blocks in the UHF bands.

Some slight inconsistencies can be seen between these figures and the recommendations given earlier. The differences are minimal and merely serve to indicate that some divergency is permissible provided the implications are considered and the planning seeks generally to avoid spectrum pollution and maximise the use of the allocated bands.

11. Separation of Private v Public Service Blocks

Whilst it can be argued that radio systems should, as far as possible, be compatible, particularly when operating together in closely knit countries, it has been found that public and private services are usually considered separately and are allocated their own blocks in which individual frequencies and channel allocations can be arranged.

This approach is based on two main factors:

(*a*)   Where both private and public systems are to be co-sited or located in close proximity on a high point, the control of potential interference is made easier if each user can be allocated channels from his own discrete block or blocks.

(*b*)   Private and public users are quite different. Public systems such as radio-telephones often require several adjacent channels with no possibility of interference.

For some kinds of user, it is often beneficial to continue this method of sub-division further by allocating small blocks of channels. This is practised in the U.K. for utility organisations. In the U.S.A. the division of users and authorities is most marked, with discrete sections allocated in quite a clear cut manner.

12. Sub-Dividing Blocks into Individual Channels

12.1 Introduction

Having allocated blocks so as to optimise isolation parameters and minimise spectrum wastage, the next step is to ensure that the blocks are used efficiently.

The overall problem of minimising the forms of interference which are likely to inhibit the correct and most efficient use of the frequency spectrum is called electromagnetic compatibility (EMC).

12.2 System Compatibility

Many users occupy sites in close proximity to others. By specifying two frequency operation, adequate isolation is assured to prevent blocking even with co-sited base stations.

This does not automatically avoid possible interference from on-channel intermodulation products, harmonics, high adjacent channel levels etc. These must be tackled on each site bearing in mind the system configurations on that site together with those of other nearby sites. Additionally, the mobile units associated with these base stations operate in a relatively large area around their own base station or stations and, whilst within this area, they must not cause or be subject to interference.

This infers that there must exist a degree of compatibility between users if an interference free environment and minimal spectrum pollution are to be achieved.

12.3 Intermodulation reduction

12.3.1 Main requirements. This subject is wide and has been adequately covered in many papers. Appendices 6, 7, 8, 9 and 10 are taken from engineering notes already published on the subject.

In frequency planning we are concerned that products (particularly low order products, third and fifth) do not fall on channels in use in the vicinity of the source of intermodulation. This does not imply the toleration of intermodulation products falling off-channel. It means that any products remaining, after a recognised minimum amount of protection has been applied, must not fall on locally-used channels.

By including such measures as are necessary to reduce intermodulation to a low level (Appendix 6) it follows that any remaining products would normally only cause interference over a relative small area, and frequency planning is therefore simplified.

12.3.2 Single communal site or conurbation of single users. Here we consider not only the normal communal site but also a conurbation of single users on individual sites within a limited area as occur in a down town area with a number of high buildings in close proximity on which individual users are located. Although physically separated, they can be considered as a single site insofar as interference is concerned.

We will discuss, in turn, the effect of intermodulation on base receivers and on mobile receivers.

(*a*)   The first category should not be troublesome. It involves conventional types of two frequency systems with each system fitting into a block plan designed to avoid low order products from the base transmitters falling in the "receiver" block. Fig. 12.1 shows a simple two block arrangement and illustrates where the main low order transmitter products will fall. The illustration, based on the typical 1 MHz blocks with 4·5 MHz spacing, shows that the first to fall into the base receiver block is the eleventh order (6a–5b) intermodulation product. This is of too high an order to cause any problems.

However, under adverse circumstances, mobile transmitters can generate or give rise to products in their own block (the base receiver block). If, for example, the base site is near enough to high concentrations of mobiles operating on nearby and related channels, then any inter-

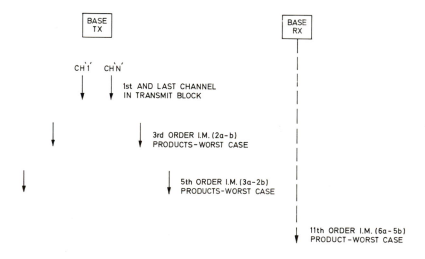

Fig. 12.1 The Effect of Base Station Transmitter Intermodulation Products on the Base Station Receiver Block

modulation product (if falling on a wanted receiver channel) will cause interference in that receiver. As can be seen, the probability is random and relies upon the combined circumstance of mobiles transmitting simultaneously on related channels at appropriate locations relative to other mobiles and the base site. Fig. 12.2 illustrates this point, based on any two mobiles radiating simultaneously. The diagram shows only the possible third order products from the four frequencies A, B, C and D. Two of these, 2C-A and 2C-D, fall on used channels (D and A, respectively). Unless they can be eliminated by removing the cause (base station receiver overload), they must be removed off-channel by re-allocation (e.g. D to E).

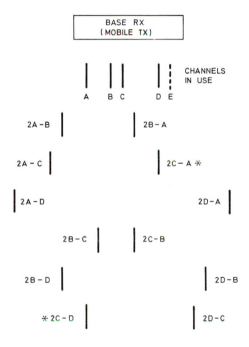

Fig. 12.2 Intermodulation Products Caused by Mobile Transmitters (any two from four radiating)

This category of intermodulation is more likely to originate in the overloading of the base receiver than be generated by appropriately positioned mobile transmitters. As stated earlier, the object of frequency planning is to prevent such products falling on a wanted channel. When such products cannot be avoided but have been minimised, then the sequence of frequencies is important if we are to place products off the wanted channels. Nevertheless, if base receivers are so overloaded by heavy concentrations of nearby mobiles that non-linearity in the base receiver causes intermodulation products to be generated, then the site location should be queried. It is not good engineering practice to locate base sites, either receiving or transmitting, in areas subjected to high concentrations of mobiles. Also, as is shown in Appendix 6, high levels of mobile signal can, if the base station is operating in the repeater mode, introduce the a + b – c form of third order intermodulation in which the base transmitter can take an active part, in spite of the base transmitter being located some megahertz from the receiver block. The process is illustrated in Fig. 12.3. The picture assumes, for both channels, a constant Tx to Rx spacing of 4·5 MHz.

Then:
$$f_b - f_a = (f_b + 4\cdot5) - (f_a + 4\cdot5)$$
So:
$$f_a - f_b + (f_b + 4\cdot5) = (f_a + 4\cdot5)$$

In other words, the combined effect of two base transmitters and a nearby mobile will give rise (in the non-linearity of Rx 1) to an intermodulation product on the same channel as the distant mobile and, maybe, of sufficient strength to prevent its reception.

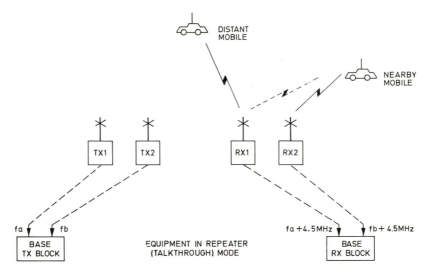

Fig. 12.3 Intermodulation Caused at Talkthrough Repeaters by High Mobile Signal Level

(*b*)   Considering mobile receivers, this category tends to be affected to a greater extent but is rather easier to pin-point and control.

Two possibilities exist. One is where the mobile receivers themselves generate intermodulation products. This is caused by the non-linearity of receiver early stages when overloaded by nearby base transmitters. As with the base receivers, the effect can be minimised by locating the base site away from high concentrations of mobiles.

The second possible source of I.M. generation, is the base transmitters themselves. As explained in Appendices 6, 7 and 9, most transmitter output stages, being Class C devices, are inherently non-linear. Therefore, owing to the close proximity of the antenna systems, coupling between the transmitter output stages can arise. Unless suitable precautions are taken, quite high levels of intermodulation can thus be generated. These products will be radiated from the antennae and can, if they fall in the mobile channels, interfere with mobiles operating within range.

The precautions to minimise this form of intermodulation are described in Appendix 10. The final step is to arrange for all transmitters within the area to be allocated frequencies on, at least, a third order freedom basis. Fifth order is much less likely but can be troublesome under adverse conditions and therefore may need special consideration.

The allocation procedure for individual sites should generally follow that described in the next section.

12.3.3 Multiple site/single user conurbation distributed over a wide area.
Assuming that intermodulation radiated from any base transmitter site is limited by suitable means (see Appendices 7 and 10) to a distance of not more than 1 km, we are left with intermodulation products generated in receivers – base or mobile. Intermodulation can occur in any receiver where two or more unwanted signals within the RF stage pass-band exceed approximately 2 mV. So, to avoid it when using 25 watt transmitters, 85 dB of isolation is required between transmitters and receivers.

Under the worst conditions, with normal antenna heights, there must be a separation of up to about 1·5 km between the *coverage perimeter* of a system and any base station site or sub-area, the frequencies of which can produce third order products falling on the channel of that system.

With this requirement, the perimeter of a mobile system likely to be affected could be near to several such sites or sub-areas. Therefore, care must be taken to ensure that any base station site or sub-area does not use transmitter channels which can cause intermodulation products to fall on the mobile receiver channel of systems where the coverage perimeter overlaps the offending base station site or sub-area. Individual sites or single-user conurbations separated by less than about 3 km should be assessed collectively for intermodulation. All those outside that distance can be considered as separate groups or areas. The distance of 3 km (1·5 km radius) has been based on a highest possible contour level of 2 mV and Fig. 12.4 shows such a configuration.

The basic principles of allocating channels having at least third order on-channel freedom are discussed at length in Appendix 8. It must be pointed out, however, that even the treatment given in this Appendix is not hard and fast. Systems within a given area, for instance, need not be allocated channels within the same band and as seen already each band can be often sub-divided into more than one pair of blocks.

Included in Appendix 9 are tables of channel combinations from a paper by Edwards, Durkin and Green (Ref. 8). These tables offer further possible degrees of sub-division.

Thus in a particular area, systems can be allocated bands best suited to them and arranged to provide freedom from at least third order on-channel intermodulation. Each area sufficiently distant from a neighbouring sub-area can employ frequencies interleaved with other sub-areas for maximum spectrum usage.

Figs. 12.5 to 12.7 show a hypothetical arrangement based on Fig. 12.4 and uses a

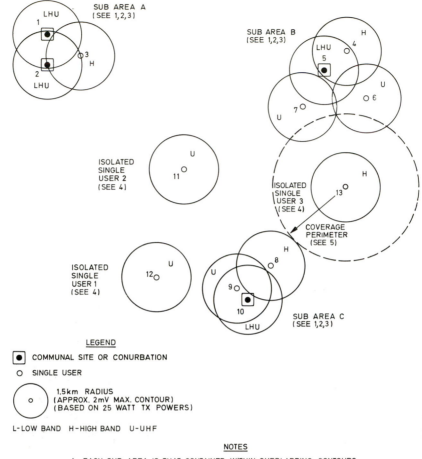

Fig. 12.4 Typical Arrangement of Sub-Areas on which to base Intermodulation-Free
Pattern

simple two block arrangement. A typical combination of channels free of third
order on-channel intermodulation has been extracted from the tables in Appendix
9. By suitable staggering, a frequency plan has been evolved covering low band,
high band and UHF systems.

Figs. 12.5 to 12.7 are based on the number of channels allocated to the total
pattern as shown in Fig. 12.8.

It is emphasised that the example is merely hypothetical to show the general
principles of channel allocation on adjacent sites for intermodulation protection.
Different channel requirements based on more elaborate block arrangements are
but an extension to this method.

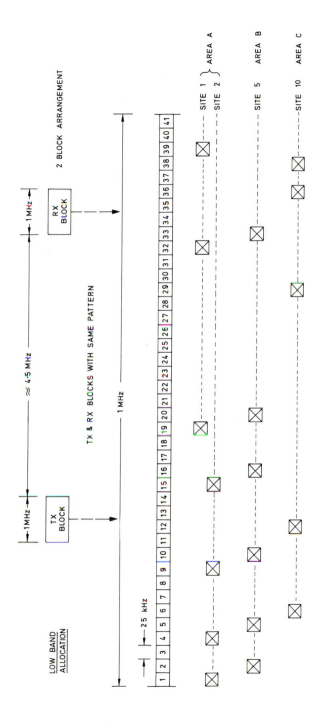

Fig. 12.5 Typical Frequency Plan for Low-Band System

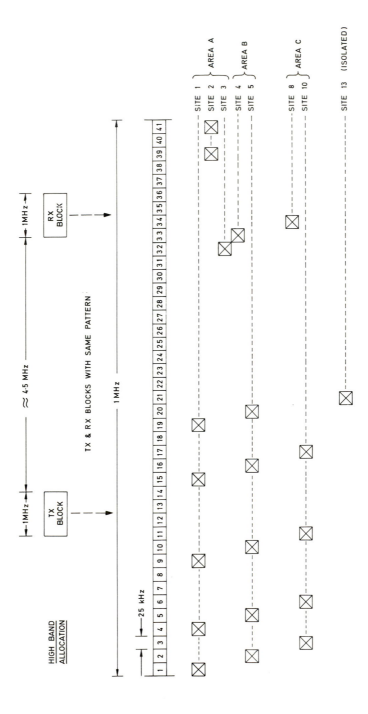

Fig. 12.6 Typical Frequency Plan for High-Band System

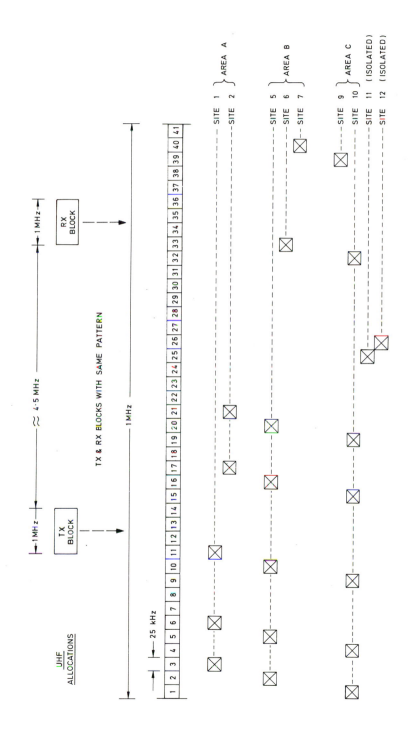

Fig. 12.7 Typical Frequency Plan for UHF System

BAND SITE	CHANNELS			
	LOW BAND	HIGH BAND	UHF	SUB AREA
1	3	5	3	A
2	4	2	2	A
3	–	1	–	A
4	–	1	–	B
5	6	5	5	B
6	–	–	1	B
7	–	–	1	B
8	–	1	–	C
9	–	–	1	C
10	5	4	6	C
11	–	–	1	ISOLATED SITE
12	–	–	1	ISOLATED SITE
13	–	1	–	ISOLATED SITE

Fig. 12.8 Hypothetical Channel Requirements of Sub-Areas Shown in Fig. 12.4

12.4   Adjacent Channel protection

Current receiver specifications contain adjacent channel selectivity requirements based occasionally on single signal measurements but more often on two signal measurements. The latter method produces figures much nearer to practical results.

Such measurements indicate that an unwanted signal removed by one channel from a wanted signal will cause a predetermined and measurable degradation of the wanted signal when its level is 70 to 80 dB above the required on-tune carrier. A complete sideband spectrum added to this figure may degrade the rejection by 10 dB or so.

Thus if the wanted signal is as low as 1 μV, the level of an unwanted adjacent channel signal would need to be approximately 1 mV, for perceptible degradation of the wanted channel. As all wanted signals in a correctly designed system are unlikely to be consistently as low as 1 μV, the average interference level on the adjacent channel could occasionally be allowed to rise above 1 mV and still remain within the 60 dB limit.

As with intermodulation, the effect at a base station depends upon the concentration of mobiles in the immediate vicinity and again it is emphasised that, base sites should be located wherever possible outside areas of intense mobile activity.

However, to minimise any such interference when allocating channels to any *one site or conurbation of single users*, the adjacent channel on either side of the wanted channel should normally be avoided in the area. Protection by distance should be included in the allocation plan and so adjacent channels should be in separate areas. Wherever possible, the chosen areas should be *outside* the normal operational radius of the allocated channel.

Figs. 12.5 to 12.7 show on-channel intermodulation free allocations complying with the above. Fig. 12.8 tabulates the allocation requirements.

Fig. 12.9 to 12.11 illustrate unwanted signal levels, noise and audio spectra in conjunction with adjacent channels and channels separated by the equivalent of a single channel.

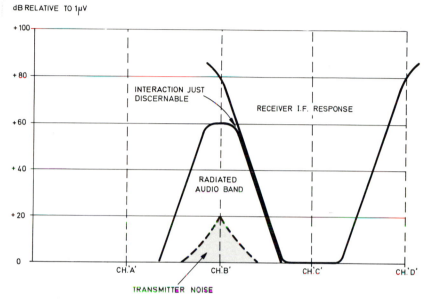

Fig. 12.9 Unwanted Signal on Adjacent Channel

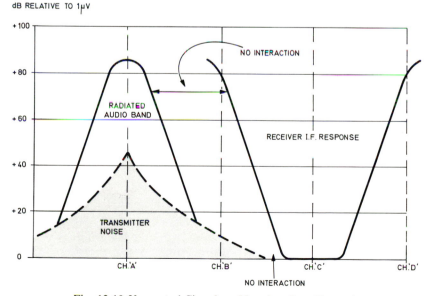

Fig. 12.10 Unwanted Signal on Next but One Channel

In Fig. 12.9, the unwanted signal is fairly small (1 mV) and on the adjacent channel, but there will be some interaction by the radiated audio, sufficient to produce just discernible degradation of the wanted signal. Blocking effects caused by non-linearity of the RF stages are not considered. The transmitter noise has no effect.

In Fig. 12.10, the unwanted signal, although now greater (20 mV), is now two channels distant and gives rise to no interaction. It is, however, at this level that blocking effects are likely to arise.

Finally, in Fig. 12.11, the unwanted signal is greater still (100 mV) and now its noise sideband interacts with the receiver, giving rise to a degraded signal-to-noise ratio.

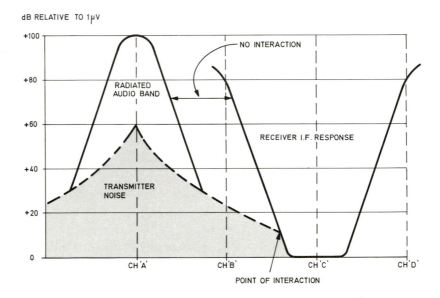

Fig. 12.11 Very Large Unwanted Signal on Next but One Channel

12.5 Harmonics and Image frequencies

Care must be taken to avoid transmitter harmonics falling on receiver channels of the same site and, if possible, in the same area.

Note that certain high band frequencies (150–174 MHz) can correspond to the second harmonic of channels in the low band (about 80 MHz). Similarly 450 MHz is the third harmonic of 150 MHz and therefore a relation can exist between frequencies in the UHF and high bands.

In theory, harmonic radiation is normally specified so as to have little effect at quite short distances from a site. However, if this site houses a number of systems, the level of harmonic radiation may be sufficient to interfere with receivers on that site, so harmonically related frequencies must be avoided. Fig. 12.12 (i) shows possible harmonic relationships with allocations in the U.K.

Similarly, although in a technically different category, image frequencies should also be avoided on the same site. For example, a receiver tuned to 168 MHz, with

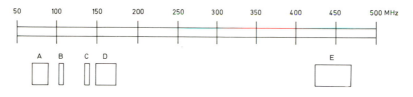

(i) HARMONIC RELATIONSHIPS IN WHOLE OR PART $\frac{C}{A} = 2$ $\frac{D}{A} = 2$ $\frac{E}{D} = 3$ $\frac{E}{A} = 6$

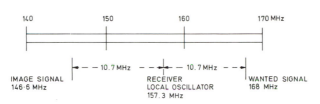

(ii) IMAGE RELATIONSHIP

Fig. 12.12 Harmonic and Image Relationships of UK allocations

the first local oscillator on the low side at 168 minus 10·7 MHz (157·3 MHz) will have an image frequency of 157·3 minus 10·7 MHz (146·6 MHz). So a local transmitter on this frequency may cause considerable interference to the 168 MHz reception. The example is illustrated by Fig. 12.12 (ii). Even though image rejection may be as high as 100 dB, a nearby 25W transmitter could, with average antenna separation, inject a signal into the receiver of approaching + 130 dB relative to 1 μV. The interference level would thus, effectively, be 30 dB above 1 μV, which is approximately 30 μV.

12.6 Split channel operation using physical distance for isolation

As the spectrum in a country or large area becomes scarce owing to high growth, obtaining additional channels by splitting the existing allocation but retaining the same basic equipment bandwidth is one way of reducing the interference potential. Such channels are occasionally called tertiary channels.

Let us examine a typical arrangement. Assuming 25 kHz channelling throughout the country and that a total of three major operational areas exist within the total area, Fig. 12.13 shows this general arrangement in a hypothetical case. If we now include a fourth area situated approximately in the centre of the original three areas (at D), it is possible that, whilst the distances between A and B, B and C and A and C suffice for re-use of all the available channels, the introduction of D will prevent any channel re-use, since unwanted signals of the order of 1 to 2 μV would be received at A, B, or C from D and vice versa.

The levels of interference between D and the other areas although only a microvolt or so, will affect all the channels involved. One obvious method of preventing

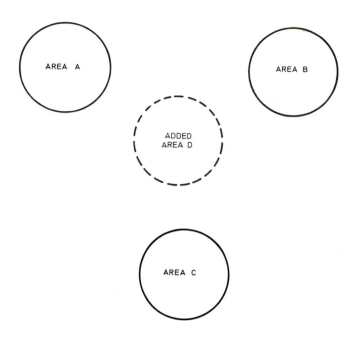

Fig. 12.13 Adding a Further Area in an Established and Planned Arrangement

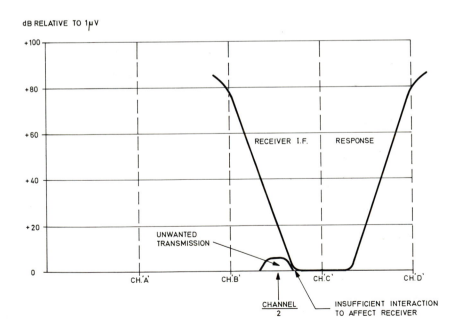

Fig. 12.14 Split Channel Operation with Standard Bandwidth Equipment

this interference from opening the squelch of the affected receivers is to employ tone squelch.

Certainly this is one solution to the problem of unwanted squelch opening, but another method has proved satisfactory in a number of countries. This method uses channels halfway between the original 25 kHz allocations but retaining the 25 kHz channelling.

Fig. 12.14 shows the relation between the unwanted transmitter radiated audio bandwidth and the receiver IF response. The interaction is negligible at the lower orders of unwanted signal level when split channel techniques are adopted. Consequently, area D can now re-use all the channels A, B and C provided they are offset by half a channel (12½ kHz).

Another advantage of this scheme is the ease of finding suitable intermodulation free channels when a large number is required. Appendix 8 explains this process in detail.

The method relies upon physical distance for attenuation; in the example, the distance being from D to either A, B or C. The unwanted signal is therefore low enough to avoid interaction with the wanted signal response.

12.7 Summarising, therefore, we note:

(a) Base station sites should not be located close to areas of high mobile activity.

(b) Transmitter intermodulation should be minimised by recognised methods.

(c) Both transmitter and receiver channels should be allocated on a third order on-channel intermodulation free basis on any one site or in conurbations of users within 3 km (fifth order freedom may need to be applied in certain isolated circumstances.

(d) The adjacent channel on either side of the wanted channel should be avoided on a single site or within a conurbation of single users and such channels should preferably be used some distance away.

(e) Harmonically related frequencies should be avoided on communal sites or in close knit areas.

(f) Split channel operation using standard bandwidth equipments can be used under certain conditions.

13. Large User Requirements

There are further requirements for users operating large systems with a number of channels.

One possible requirement is the containment of the channels allocated to a large user in one band to within a certain spread. This enables a single equipment to switch between channels in the normal way without re-tuning or requiring additional RF receiver heads. This requirement, together with those previously described, may be served by setting aside a special section of the transmit and receive blocks for the purpose. Indeed, with large specialised users, such as public utilities, ambulance and fire services, this is often achieved by allocating two distinct blocks (transmit and receive) for them alone, possibly combining similar users in a special block arrangement.

Police are usually catered for by a special pair of blocks. This is particularly so in a close knit countrywide arrangement for emergency use but where each basic sub-area operates separately on a normal day to day basis.

The problem is often solved by such large users occupying exclusive sites isolated from those of other users.

These users often need extensive coverage and therefore such specialised sites are generally located on prominent hilltops in rural areas. Consequently, a high concentration of nearby mobiles is extremely unlikely and intermodulation and other interference problems are therefore somewhat easier to avoid.

Limited coverage is sometimes needed by such users. For example, a small portion of a city by a police division; this would be achieved by a local UHF system at the police station. These systems usually operate only to portable equipments so, again, many of the problems associated with multiple user sites and high power mobiles do not exist. One should however obey the general principles of frequency planning to prevent subsequent problems if the mode of operation is modified. Where a concentration of relatively low power stations exists, the radius postulated in Section 12.3.3 can be reduced. Typically, an interference radius of 1·5 km would be reduced to about 0·75 km if the transmitter power were reduced from 25 watts to 5 watts.

14. Spectrum Economy by Range Restriction

Spectrum use concerns not only the *bandwidth* occupied by a transmitter but also the *area* in which this portion of the spectrum is used. A further "dimension" can also be directly related to the *time* the segment of spectrum is used. (Appendix 4)

We can control the bandwidth of the systems to which we allocate frequencies by appropriate specification. Insofar as time is concerned, we can endeavour to control the occupancy of a channel so as to optimise its usage.

The remaining "dimensions" (area) involves two points:

(*a*)   The coverage of the system.
(*b*)   The distance necessary between sites before the channel can be re-used without interference.

Both these are contradictory but illustrate clearly that excessive or unnecessary coverage is wasteful of spectrum.

At 150 MHz, operational range in km approximates to

$$4 \cdot 1 \, (\sqrt{h_1} \; + \; \sqrt{h_2})$$

Where h_1 = base antenna height in metres relative to average terrain height.
h_2 = mobile antenna height in metres.

and it can be seen that the dominant dimension is the base station antenna height.

Appendix 2 discusses re-use distances at length. It is seen that channel re-use is possible when the distance between the two sites using the same frequency is approximately equal to the sum of the smaller operational range and four times the larger operational range. For example, if the operational ranges from two sites are 18 and 25 km respectively, the re-use distance will be approximately 118 km.

Analysing the requirements of these two systems it may well be that their operational ranges need to be substantially the same, perhaps the smaller value (18 km). Then unless there is an obstruction which must be overcome by the extra effective height, there is no reason why the height of the base antenna providing the

25 km range should not be reduced. If both operational ranges are 18 km, then the re-use range is lowered to 90 km.

Transmitter power, particularly that of the base transmitter, affects range to a lesser degree. Transmitter power influences the received signal to noise ratio dB for dB. So four times the power (6 dB) will improve the signal to noise by 6 dB, thus possibly improving reception in an electrically noisy area within the coverage range.

Transmitter power also similarly offsets propagation loss but as propagation loss increases rapidly as the radio horizon is approached, the gain in actual distance is relatively small unless the power increase is considerable. Thus, if range is more than adequate with a high order of signal to noise performance within the operational area, then reducing base station power could be considered. This might give lower interference within the area and make adequate isolation between systems easier to achieve.

This does not mean that system performance must be reduced to the bare minimum. Rather, it is intended to indicate that high powers and high sites are not always necessary. In fact, under certain circumstances, two separately located channels each with reasonable antenna height and power, can often show greater spectrum conservation than a single channel operating at high power into an antenna located on a high hill or mountain. Both in total may cover the same *required* area but the *unwanted* coverage of the single channel high sited system may be vastly greater than needed. In mountainous areas, the arrangement with lower antennae is often cheaper by eliminating the need for an isolated or perhaps inaccessible high site. Often the lower sites can, owing to nature's screening, use the same channel and this can be re-used more frequently than is normally possible over flat terrain. Fig. 14.1 shows such a scheme with three possible sites, A, B (both low) and the high site shown with coverage radii of 19, 14 and 46 km. The re-use distances are, therefore, 95, 70 and 230 km, respectively. Thus, for coverage of areas A and B, the use of the two low sites A and B is far more conservative of spectrum.

Range may need to be restricted to avoid radiation in certain areas. They might, typically, be sea areas, neighbouring countries or mountainous regions. Here directional antennae are often useful. As used in radio links, they provide the equivalent of a transmitter power increase in the direction(s) of high gain at the expense of a reduction in the direction(s) of minimum gain.

For example, an antenna gain of 6 dB is equivalent to a fourfold increase in transmitter power. Thus, such antennae, often permit lower power base transmitters and consequently improve other characteristics.

A similar advantage may be gained by using a directional gain antenna connected to a base receiver to increase the signal level from the mobile transmitter in a particular area.

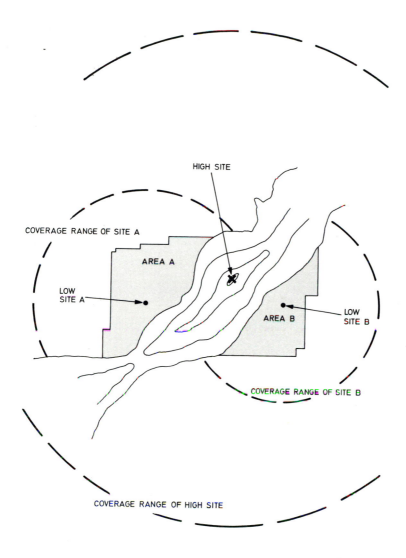

Fig. 14.1 Effect of High and Low Sites on Channel Re-use

15.   Channel Occupancy

15.1   General considerations

Here we are concerned with the amount of traffic permitted over the channel before it is considered to be overloaded and thus inefficiently utilised.

Channel overloading produces user frustration, long delays, loss of message value and (on communal sites) excessive interference.

Although widely discussed, it is a poorly understood problem. Appendix 5 discusses channel occupancy at length giving the necessary procedure and showing the implications of overloading.

Suitable statistics are pitifully short. Information such as average message length, evaluation of the busy hour, average call rate and the number of mobiles in use during the busy period are all essential to the assessment of channel occupancy. Much of this detail can be obtained only by suitable monitoring or from information derived from experience (Refs. 7, 10, 12 and 17).

Each type of user has individual characteristics and the art of sharing channels hinges on the correct analysis of these characteristics.

The busy period, usually a period of one hour taken out of the total 24 hours, is analysed and the percentage of this period during which the *channel* (both go and return paths) is occupied, indicates the loading of that channel.

Where channels are shared, and if the different users occupy the channel in different ways, greater channel utilisation can be achieved. For example, users' busy hours may not coincide.

Appendix 1 gives a broad view of channel sharing and should be consulted with Appendix 5 for a general picture of channel occupancy.

Summarising, the busy hour occupancy of a radio *channel* should not exceed 50% if excessive waiting times are to be avoided. In services of literally vital importance, the loss of message value with increased delay can be crucial, and a maximum loading of 40% may be preferable.

To calculate the loading of a single channel during the busy hour is quite simple. It can be expressed as

$$\frac{t\,v\,c}{36}\;\%$$

Where t = average effective message length in seconds
     v = average number of vehicles passing messages during the busy hour
     c = average number of messages per vehicle during the busy hour (call rate).

Let us, therefore, as a typical example take

    t = 20 seconds
    v = 30 vehicles
    c = 2 messages

Channel occupancy with these values is 33·3%, well within the 40 to 50% postulated as a reasonable maximum.

Obviously, users with much shorter messages, such as taxis, would show a considerably lower individual loading so more vehicles could be employed for the same occupancy.

Evidently, a knowledge of channel occupancy is essential for a correct assessment of the number of channels to be allocated to the different types of user. Furthermore it is essential for correct shared usage of a channel to avoid excessive peaks and consequent frustration at the times of maximum demand.

As well as ensuring peak hour staggering in a shared system, the geographical user locations are important for maximum re-use. Appendix 1 deals with this subject in detail.

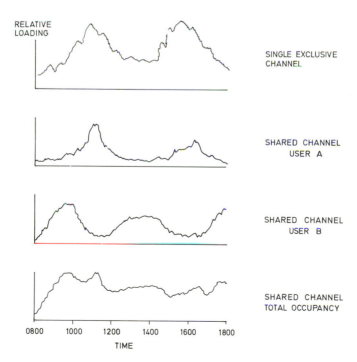

Fig. 15.1 Shared Use of Mobile Radio Channels

From the overall treatment, we see that, when properly organised, a channel shared by a number of small users can be more efficiently occupied than by one large user only. So more mobiles can be specified per channel in the multi-user case. This arises in spite of probably better operating technique in the single case. Clearly, the level of occupancy should be smoothed as much as possible over the working period. Fig. 15.1 gives a hypothetical example.

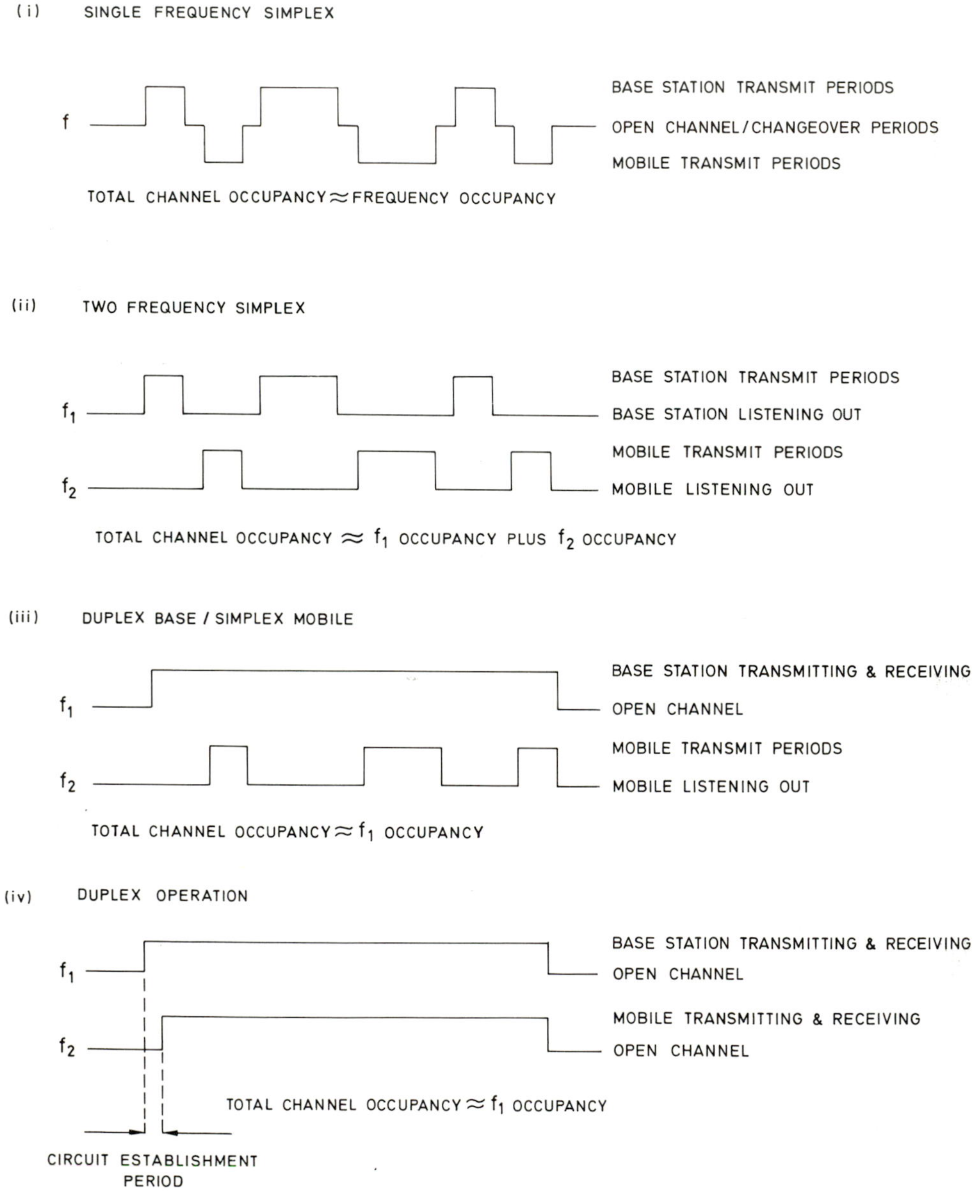

(i)   SINGLE FREQUENCY SIMPLEX

f BASE STATION TRANSMIT PERIODS

                                            OPEN CHANNEL/CHANGEOVER PERIODS

                                            MOBILE TRANSMIT PERIODS

     TOTAL CHANNEL OCCUPANCY ≈ FREQUENCY OCCUPANCY

(ii)   TWO FREQUENCY SIMPLEX

f_1 BASE STATION TRANSMIT PERIODS

                                            BASE STATION LISTENING OUT

                                            MOBILE TRANSMIT PERIODS

f_2 MOBILE LISTENING OUT

     TOTAL CHANNEL OCCUPANCY ≈ f_1 OCCUPANCY PLUS f_2 OCCUPANCY

(iii) DUPLEX BASE / SIMPLEX MOBILE

f_1 BASE STATION TRANSMITTING & RECEIVING

                                            OPEN CHANNEL

                                            MOBILE TRANSMIT PERIODS

f_2 MOBILE LISTENING OUT

     TOTAL CHANNEL OCCUPANCY ≈ f_1 OCCUPANCY

(iv)   DUPLEX OPERATION

f_1 BASE STATION TRANSMITTING & RECEIVING

                                            OPEN CHANNEL

                                            MOBILE TRANSMITTING & RECEIVING

f_2 OPEN CHANNEL

          TOTAL CHANNEL OCCUPANCY ≈ f_1 OCCUPANCY

   CIRCUIT ESTABLISHMENT
   PERIOD

Fig. 15.2 Various Modes of Operation

15.2 Effect of single frequency and two frequency operation

The channel occupancy is the fraction of a predetermined period during which the *channel* is occupied by one or more users. Considering the transmission mode, such as single or two frequency simplex, differences can be seen in the actual occupancy of the *frequencies* involved.

Fig. 15.2 (i) to (iv) shows the various modes of operation in a simple base station originated call.

In the single frequency simplex case, the one frequency involved is used for both the go and return paths and consequently the term "channel" also applies to this single frequency. During a complete conversation the channel is fully occupied – with the exception of the very short changeover pauses – and therefore the channel loading factor is also the loading factor of the *frequency* used.

In the two frequency simplex case, the two frequencies are used alternately. Although the channel occupancy is the same as in the single frequency case each frequency is loaded with perhaps only 50% of the total traffic. There is, of course, again a slight pause at each changeover point.

The third example shows the duplex base/simplex mobile case. Here the frequency occupancy of the base transmission is equal to the channel occupancy; the mobile frequency occupancy is approximately 50% of the channel occupancy.

Finally, in the full duplex case, apart from the slight period necessary for the establishment of the circuit, both the base and mobile frequency occupancies are equal to the channel occupancy.

16. The Use of Tone Squelch in Shared Systems

Where a number of users, each with low density traffic, occupy a common channel within a limited area, the need for operational discipline becomes more important with increased channel loading when additional mobiles or further users are added.

The effect of high levels of loading in multi-user common area shared channels is described in Appendix 1, particularly in relation to obliteration. Whilst the likelihood of two or more mobiles simultaneously radiating cannot be eliminated, means exist to reduce the annoyance of listening continuously to other users.

The usual solution is tone squelch in which the receiver squelch circuits are opened only by a carrier modulated by a particular tone (below 300 Hz). If each user employs a different tone, then only that tone will open the channel for that user. Unfortunately, once opened, other users will also be heard. The method does, at least, ensure that the fixed and mobile receivers on any one net are quietened during the periods that the user's transmitters are not operating.

Appendix 17 describes at length the system principles.

17. Operation of Mobile Radio Systems Over Radio Links

17.1 Method of control

As explained earlier (14), the coverage of a mobile radio system is a function of antenna height. It is therefore often necessary to elevate an antenna system by hundreds rather than tens of metres for adequate coverage over the wide area needed by certain types of system.

One convenient method in some types of terrain is to use a prominent hill or mountain, if one can be found overlooking the required coverage areas. Unfortunately, this high site is often some distance from the point of control needing either landlines or a radio link between the two locations.

The choice of a landline or a radio link will depend upon the total requirements as well as availability.

Generally, a single landline pair of a certain minimum standard will be suitable for the control of a single standard base station. Additional facilities are often an extra complication which may necessitate a radio link. Obviously, if a landline does not exist, the cost of a radio link is usually far less than that of laying a cable. A direct radio link may occasionally not be feasible in the absence of a suitable propagation path. Series links are then needed.

We therefore concentrate on the radio link solution and analyse the main requirements.

17.2 Mode of Operation

An important decision is whether to use single frequency or two frequency operation. As with mobile radio, it has been proved that spectrum saving is associated with two frequency operation. On a central site, from which more than one link channel radiates, single frequency operation would be extremely wasteful of spectrum.

Additionally, single frequency operation on a link system can result in lock out and Fig. 17.1 illustrates how this occurs.

Fig. 17.2 shows how, with two frequency operation and a suitable antenna arrangement at the remote base station site, lockout of the system can be avoided. This permits full use of the channel without the wastage which would occur during periods of lockout.

17.3 Limitation of effective radiated power

In the interests of frequency conservation it is essential to limit the radiated power from link transmitters. Owing to high sites, the radiated signal can extend over a wide area if not restricted.

The main method used in this context is to fit high gain antennae which have narrow beam widths in the horizontal plane. This concentrates the radiation to the required direction and reduces it in unwanted directions. A similar arrangement at the receiving terminal increases the system gain still further whilst the directional properties of that antenna minimise the pickup from all directions except that of the required transmitter.

The transmitter power can now be correspondingly reduced.

The design values used in the U.K. are shown in Fig. 17.3. They are eminently suitable for the single channel link needed for controlling mobile radio base stations. It is recommended that similar criteria be adopted for any new planning arrangement.

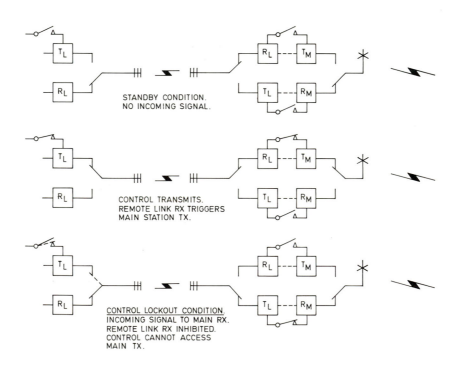

Fig. 17.1 Single Frequency Simplex Link Operation

(i) TWO FREQUENCY SIMPLEX OPERATION

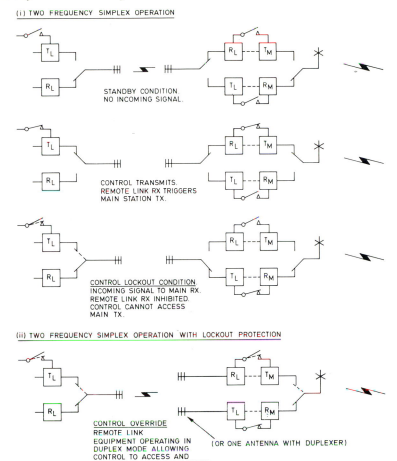

Fig. 17.2 Two Frequency Simplex Link Operation

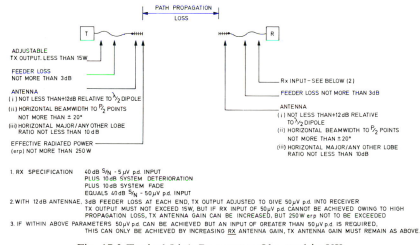

Fig. 17.3 Typical Link Parameters Observed in UK

18.   Frequency Planning of Radio Link Systems

18.1   General considerations

Link systems are, in many respects, much easier to plan than mobile radio systems. On the other hand there are one or two important points to consider which, in mobile radio would not necessarily be considered in quite the same light.

18.2   Block configuration

It is essential, for reasons shown later, to allocate at least *two* pairs of blocks in each band reserved for point to point use. The spacing between the transmitter and receiver blocks should obey the principles given in Section 10 but, in addition, the pairs of blocks should also be separated by an amount equal to the transmitter to receiver spacing. Fig. 18.1 shows a simple typical arrangement and it can be seen that by spacing the blocks in the manner required, not only can blocks 1 and 2 or blocks 3 and 4 be used when allocating channels, but in certain circumstances block 2 can be paired with block 3 while keeping the desired isolation. Furthermore it can be seen that a fourth block pairing is also possible, between 1 and 4. This last arrangement, as can be seen, will have a much greater frequency separation.

Note that, to avoid transmitter intermodulation products falling in receive blocks, it is essential to avoid allocating adjacent blocks for the transmit function on the *same* site.

Fig. 18.1 A Simple Typical Radio Link Block Allocation pattern

18.3   Intermodulation potential

Link systems are, in most cases, less susceptible to the worst effects of intermodulation than mobile radio systems. This is mainly owing to the fixed nature of links compared with mobile systems where the mobile calling could be anywhere between the coverage area perimeter and a few metres of the base site. In these circumstances the mobile signal into the base receiver could vary by as much as 60 to 80 dB with the reverse path from the fixed transmitter to the mobile receiver possibly exhibiting an even greater spread.

In comparison, a link system path would normally only change during fades. Depending upon the type of path, terrain, climate etc., the signal would probably only vary by between 10 and 40 dB. The lower ratios are most likely in typical systems.

Examining Fig. 18.1 it can be seen that transmitter intermodulation would not fall in the receiver blocks in the example given – at least, if one considers only the low order troublesome products. Additionally with the levels of the received signal likely from distant sites, the generation of intermodulation by non-linearity in the receivers is unlikely. However, if the level of these signals is higher than needed (hundreds of micro-volts), then by the a + b – c effect, some unwanted products may be generated in conjunction with the link transmitters located on the sites. A workable level of signal should be used – neither too high nor too low. The requirement for a suitable signal level has been discussed for other reasons in Section 17.2 and applies equally well in this situation.

Although intermodulation is less obtrusive in correctly designed radio link systems, precautions should however be taken to minimise generation of products in order to ensure that spectrum pollution is of a low level. Additionally, in frequency planning, third order on-channel intermodulation free allocations should be pursued, as with mobile radio, for optimum system design.

18.4 Frequency planning procedure. Block allocations

In allocating channels to a number of radio link systems within an area it is important to begin at the site housing the largest number of link terminals. If a start is made at a single terminal site, it will often be found that, upon arrival at the larger sites, incompatibility in block use arises or channel utilisation will be poor. Problems such as interference will also be greatest at the more congested sites.

As with allocation for mobile systems (see 10.2), it is most important to arrange that *all* transmitters on any site are separated from *all* receivers on that site by a frequency spacing sufficient to provide the desired isolation.

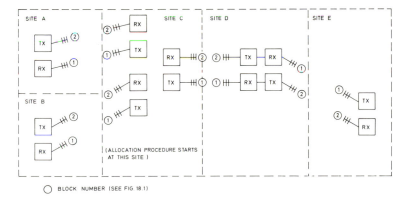

Fig. 18.2 Typical Frequency Planning for Link Terminals

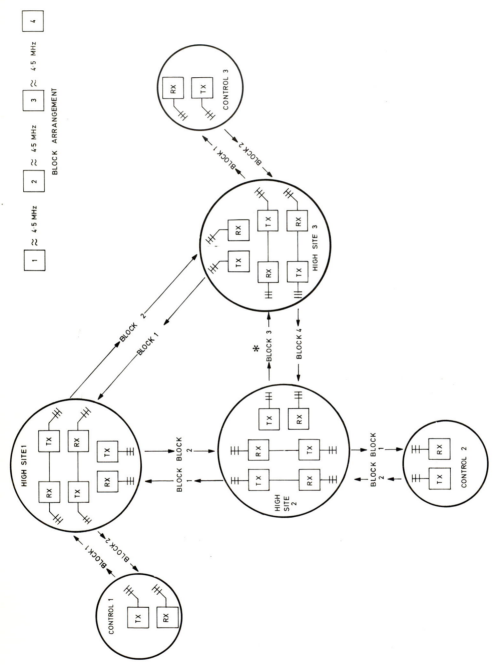

Fig. 18.3 (Hypothetical) Example Showing Need for 2nd Pair of Blocks with Full Inter-block Spacing

These requirements should therefore be followed for the correct use of the blocks shown in Fig. 18.1.

If we now examine Fig. 18.2 in which a relatively simple arrangement is shown, we can see that the site having the greatest number of link terminals is site C. We therefore start at this point and allocate channels in block 1 to all transmitters and similarly channels in block 2 to all receivers. We now transfer these channels to the related equipments at the distant terminals and then continue to work outwards until the allocations are complete, always ensuring that on any one site all the transmitters and receivers are sufficiently isolated in frequency. Where incompatibility exists as shown in Fig. 18.3 (*between high sites 2 and 3) use must be made of the second pair of blocks. Incompatibility in this case is caused by the clash in the use of blocks 1 and 2. Transmitters on high site 2 are shown in block 1 in two cases, but the pattern would need to be reversed with a receiver in block 1 if the overall trend were continued towards high site 3. This would cause severe blocking on high site 2 of all receivers and therefore another block sequence such as blocks 3 and 4 must be used for the connecting link (high site 2 to high site 3).

A quick means to check compatibility is to plot the area path pattern as shown in Fig. 18.4. If between the points where the paths separate and converge again, the

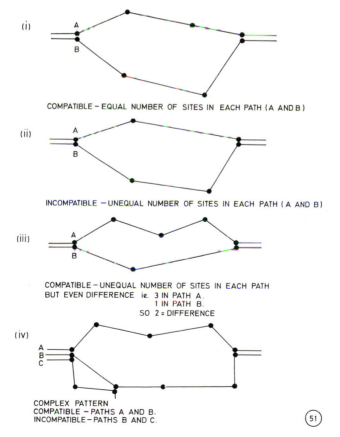

Fig. 18.4 A Method of Checking Frequency Compatibility

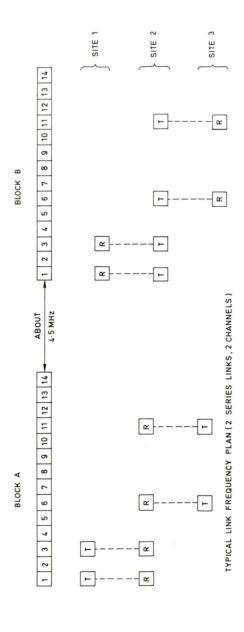

Fig. 18.5 Typical Link Frequency Plan

number of repeaters or sites in all the paths are equal or have even differences then compatibility exists. If they have odd differences, separate block pairs are needed.

18.5 Individual channel allocations based on on-channel intermodulation freedom

Since the procedure follows very closely that proposed for mobile radio planning, the detail is not repeated. Fig. 18.5 shows a typical site allocation necessary for a radio link system. As with the mobile procedure, the link blocks are divided into channels, 25 kHz in this case.

Based on a suitable third order on-channel intermodulation free pattern, the channels are allocated to suit the particular site. Other sites in the area are treated in a similar manner but with the channels suitably staggered from those already allocated.

The example shown by Fig. 18.5 is just one possible arrangement using a typical third order on-channel intermodulation free sequence. It will be noted that this sequence has been used both in a forward and reverse direction to obtain maximum utilisation of the blocks.

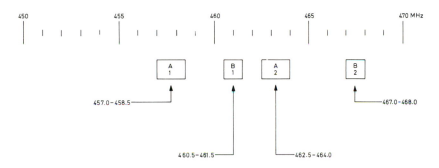

Fig. 18.6 Available Spectrum in the U.K. for Private UHF Point-to-Point Links (1977)

18.6 Example of United Kingdom UHF pattern

Fig. 18.6 shows the point to point block pattern at present in use in the United Kingdom. It follows the above principles although it will be seen that there is a reduced spacing between B1 and A2. This is an unavoidable restriction, but in the majority of allocations it can be used without additional filtering. Where extra filtering is required, it is within technical limits as the frequency spacing, although less than desired, is still sufficient to provide a reasonable portion of the isolation needed.

It is anticipated that, in the next decade, the UHF band will be cleared of these point to point links which will be moved to higher bands (e.g. 1500 MHz), thus vacating the 450 MHz band for additional mobile radio use.

18.7   The allocation of channels to point to point links in bands above 1700 MHz

Allocation of channels in these higher bands, particularly for wide band links such as multiplex or music bearers, fall into a different category from the mobile radio single channel plans. Hence they usually have a predetermined pattern which has been internationally agreed by C.C.I.R. This makes due allowance for the necessary bandwidths and adjacent channel protection and should be followed in any allocations made in the bands so affected. This is especially necessary for links which are part of an international system or those near to national boundaries.

The most decisive aspect in frequency planning in the bands above 1·7 GHz as practised by C.C.I.R. is, of course, frequency economy using state of the art aids such as antenna directivity and filters. The frequencies allocated to radio relay stations should also be arranged so that several RF channels can be operated in parallel within one hop. The same antenna should preferably be associated with several RF channels. It should be possible for radio paths to cross with minimum interaction.

In addition to the above, standard partial frequency ranges laid down in international assignment patterns (otherwise described as special national or supernational assignment patterns) are available as a result of co-operation between administrations and equipment manufacturers.

19. Acceptable System Degradation Limits

19.1 Limiting factors

The efficiency of any communication system depends upon the combination of many factors. These factors can be broadly divided into two categories:

(*a*) Those which are fundamentally limiting.

(*b*) Those which are a function of the system or its environment and can be improved or eliminated.

The first category includes such factors as basic signal to noise performance of a receiver, the inherent non-linearity of a receiver or transmitter and the many other factors which either are fundamental or cannot be designed out in an economic manner.

In the second category we can include, for instance, local site noise, inter-modulation products, acoustic and electrical noise in the vehicle and adjacent channel interference. We will concentrate on the comparative assessment of those effects rather than on cures or palliatives.

19.2 Assessing degradation

External noise, caused, for instance, by the electrical systems of nearby vehicles constitutes, theoretically, a fault condition, but it cannot normally be reduced to an acceptable level without the introduction of laws imposing levels of mandatory suppression. Such laws have already been introduced in many countries but, in those countries without such legislation, the effect is often a marked reduction in the effectiveness of each radio system. Levels of electrical noise, including vehicle impulsive noise, are usually quite high at the lower frequencies (40 MHz) often reaching levels equivalent to $+30$ dB relative to 1 μV in city centres. In rural areas the levels are about 20 dB lower.

As frequencies are increased, the man-made noise level is lower. The noise at 450 MHz is about 6 dB relative to 1 μV in city areas and practically non-existent in rural areas.

Consequently the choice of frequency band is important with noise levels and propagation efficiency competing. (See Section 7 and, in particular Fig. 7.1

showing the noise levels as given by C.C.I.R.)

Let us now consider levels of interference from sources other than "electrical". By this we mean interference from co-channel, adjacent channel, intermodulation, cross modulation and blocking effects. Here, however, we are concerned with fault conditions involving other users of the radio spectrum and in theory it should be possible to eliminate the effect. In practice, unfortunately, we shall probably be left with some finite level of interference, whether in the form of noise or speech, intelligible or non-intelligible.

In all cases however we are concerned with obtaining an adequate standard of communication.

Let us assume, therefore that for acceptable intelligibility a signal to noise ratio of 20 dB is needed. If this figure is obtainable with an input of, say 2 μV into a receiver operating under ideal conditions, then it is obvious that if unwanted noise is introduced, to maintain the desired 20 dB, the wanted signal input must be increased. The increase needed indicates the severity of the interference.

The degree of interference can also be assessed from the reduction in signal to noise ratio with constant input and this method is often preferred and therefore generally adopted in many receiver bench measurements.

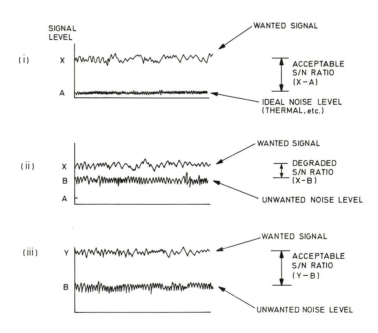

Fig. 19.1 Signal Degradation by Noise

19.3 Summary

We can assess in the field the amount of unwanted noise etc. by determining the additional level of wanted signal needed, relative to an ideal case, to provide the wanted acceptable signal to noise ratio (20 dB for example). Bench or site measurements to determine the acceptable degree of interference can also be based on a pre-determined degradation of the wanted signal. This level of degradation of the wanted signal is usually accepted as 3 dB, although, in some cases, 6 dB is preferred. Fig. 19.1 shows how noise can degrade the wanted signal. In (i) it can be seen that a low level of noise at A permits the wanted signal to be located at X to provide an acceptable signal/noise ratio.

In (ii) the increase in noise from A to B has worsened the signal/noise ratio by an amount equal to the difference between A and B.

To restore the original acceptable signal/noise ratio with the higher level of noise, it is necessary (iii) to increase the level of wanted signal to Y.

20. Equipment Specifications

This is important because accurate frequency planning depends upon a proper knowledge of the fixed parameters. Variable parameters, often quite numerous, make the procedures uncertain. Hence margins have to be increased or the planning has to be otherwise improved.

Equipment specifications can be hazardous. Limits should be technologically reasonable and not so high as to require a special non-commercial design. Limits given to conflicting parameters should be carefully chosen to avoid the condition where neither can be met. Those parameters not directly bearing on interference or frequency planning may not need limits.

One solution is to use an established specification, such as that applied to equipment used in the U.K. Alternatively only the important parameters should be given limits. Amongst these parameters are:

(a) *Frequency Stability*
   The frequency stability required of all transmitters and receivers, both base and mobile, should be in keeping with the channel bandwidth adopted, the frequency band involved, and the ambient temperature range of the country concerned.

(b) *Spurious outputs*
   The level of spurious outputs, from any transmitter and to the limited degree possible, from any receiver, should be set at a level based on current design and practice.

(c) *Transmitter output power*
   This is one of the more debatable areas. Some countries have a high top limit, others quite low. A maximum of around 25 watts is accepted by a number of countries with a lower figure at UHF where high gain antennae are often used.

(d) *Bandwidths*
   These involve the receiver's limit of acceptable adjacent channel rejection and the transmitter radiated bandwidth limits. Both must be compatible; the latter should specify the peak level of modulation permissible.

(e) *Receiver spurious response performance*

Spurious responses can affect the performance of a receiver on a communal site.

(*f*)   *Receiver intermodulation performance*
This is basically an extension of (*e*) above in which the level of unwanted signals to produce intermodulation products determines the limit imposed.

Many countries already have agreed limits for the above parameters and, consequently, many commercial equipments have been designed with these limits in mind. Tighter limits are often unnecessary, occasionally impossible and usually prohibitive in established designs.

21. Records

21.1 General considerations

Apart from the actual procedure of frequency planning the keeping of records is undoubtedly the most important step.

Whether sophisticated modern computer techniques are employed or the simple paper register entry is used, it is essential that the record of every allocation is complete to the smallest detail. The few extra moments taken at the planning stage to ensure that adequate detail is available can save hours at a later date; the completeness of the detail will undoubtedly lead to better planning; whilst the cost of errors brought about by insufficient or incorrect information can be extremely high.

One must always remember that the main objective of frequency planning is spectrum conservation and, as a continuous process, each step hinges on the previous result. Unless all details of the previous steps are retained in a concise and readily available form, the planning process will eventually slow down and errors will arise.

It is therefore important to examine in some detail, the type of information which should be included in the frequency planning records. Such information could fall into the following typical categories:

21.2 Application Reference Number

21.3 User detail

(a) Name
(b) Full address
(c) Telephone number(s)
(d) Contact personnel and home telephone numbers
(e) Type of business
(f) Any other relevant details

21.4 Brief System detail

(a) Number of base stations

        – Manufacturer
        – Type number
(b) Number of controls
        – Landline operation
        – Radio link operation
        – Reverse frequency talkthrough
(c) Number of mobiles
        – Manufacturer
        – Type number
(d) Number of Portables
        – Manufacturer
        – Type number
(e) Number of fixed outstations
        – Manufacturer
        – Type number
(f) Number of radio link paths, if used
(g) Any other relevant details

21.5 Base Station Site detail

(a) Location and address
        – Map references etc.
(b) Height above sea level
(c) Mast detail
        – Type
        – Total height
        – Height of antennae
(d) Transmitter output power
(e) Antenna system
        – Type of radiation pattern
        – Gain characteristics
(f) System Channel Bandwidth
(g) Modulation method
        – AM
        – FM/PM
(h) Mode of operation
        – Single frequency simplex
        – Two frequency simplex
        – Duplex (intermittent)
        – Duplex (continuous)
(i) Talkthrough facility
        – Fitted or not
        – Switched over line
           – d.c. switched
           – tone switched
        – Switched over radio link
        – Permanently on talkthrough
        – Through complete link path/control

(*j*)   Tone systems
    – Fitted or not
    – Selective calling
       – method/type
    – Tone squelch
       – tone frequency used
(*k*)   Transmitter Duty Cycle (estimated)

21.6   Control Site(s) details

(*a*)   Location and address
    – Map references etc.

21.7   Base Station Traffic detail

(*a*)   Normal hours of use
    – Daytime
    – Night
(*b*)   Average message length
(*c*)   Calls per mobile
    – 24 hours
    – busy hours
(*d*)   Busy hour loading estimate
(*e*)   Any other relevant detail

21.8   Channel(s) allocation – Base Station

(*a*)   Base station(s) Tx   Rx
(*b*)   Special application Tx/Rx
    – single frequency mobile to mobile

21.9   Channel(s) allocation – Radio Link

(*a*)   Terminal A   Tx   Rx
(*b*)   Terminal B   Tx   Rx
(*c*)   Any repeaters   Tx 1   Rx 1
                        Tx 2   Rx 2

21.10   Details of radio link(s) if fitted

(*a*)   Number of link paths
(*b*)   Location and address of Terminal A
    – Map references etc.
(*c*)   Location and address of Terminal B
    – Map references etc.
(*d*)   Locations and addresses of any repeaters
    – Map references etc.
(*e*)   Height above sea level of Terminal A
(*f*)   Height above sea level of Terminal B
(*g*)   Heights above sea level of any repeaters
(*h*)   Mast detail at Terminal A
    – Type

– Total height
– Height of antennae
(*i*)   Mast detail at Terminal B
– Type
– Total height
– Height of antennae
(*j*)   Mast details of any Repeater
– Type
– Total height
– Height of antennae
(*k*)   Transmitter output power at Terminal A
(*l*)   Transmitter output power at Terminal B
(*m*)   Transmitter output power at any Repeaters
(*n*)   Antenna systems at Terminal A
– Type of radiation pattern
– Gain characteristics
(*o*)   Antenna systems at Terminal B
– Type of radiation pattern
– Gain characteristics
(*p*)   Antenna systems at any Repeaters
– Type of radiation pattern
– Gain characteristics
(*q*)   Link Channel bandwidth
(*r*)   Modulation method
(*s*)   Mode of operation
– two frequency simplex
– duplex (intermittent)
– duplex (continuous)
(*t*)   Tone system detail if fitted

No.	SITE DETAIL – ADDRESS / LOCATION	MAP REFERENCE	CHANNELS ALLOCATED	USERS – APPLICATION REF. No.	
1.	SHOOTERS HILL – STREATHAM	FA 434 271	C21	BROWN C⁰	27
			C39	SMITH	91
			D4	JONES	2
			E105	MACTAVISH	107
			E109	PONTIN	109
				ETC.	
2.	HAGNELLS POINT – MANDEVILLE	QT 222175	A1	AAA TAXI	99
			A11	R.D. CEMENT	12
			A19	T. RULE	82
			D5	GENSON	205
			C62	GARDEN PRODUCTS	209
			E47	THOMPSON	65
			E99	PREVITT	59

Fig. 21.1 Site Cross Reference – Example

21.11 Cross references

Although the basic information can be tabulated in a manner similar to that given above, this information should also be cross-referenced to the two main parameters, site and channels allocated.

Thus we have

(*a*)   User (reference number)
     Basic information 21.2 to 21.10
(*b*)   Site (see Fig. 21.1)
     Cross reference to
     (i)   Channels allocated to site
     (ii) Users of site (reference numbers)
     Note: It may be preferable to allocate one sheet to each site and file alphabetically.
(*c*)   Channels allocated (see Fig. 21.2)
     Cross reference to
     (i)   Site
     (ii) User (reference number)

CHANNEL REF.	TX FREQ.	RX FREQ.	SITE DETAIL		USERS	APPLICATION REF. No.
			No.	LOCATION		
A1	76·500	81·000	6	WRIGHTS HILL	THOMAS	23
A2	76·525	81·025	12	JAMAICA POINT	PINCHIN	3
A3	76·550	81·050	4	MOKATTAM MT.	EL KADI	94
A4						
A5						
A6						
A7						
A8						
A9						

Fig. 21.2 Channel Cross Reference – Example

21.12 Use of Channel numbering system

In a number of countries, it has been found advantageous to allocate an actual code number to a channel (comprising the transmit and receive frequencies) in preference to referring to the actual frequencies.

This is of advantage both in record maintenance and in the greater ease of identifying intermodulation products. Each band can use the same numbering systems but with a letter of the alphabet prefixed to identify the band used. The prefix can also be used to indicate the block pair where more than a single pair is used in a band.

Fig. 21.3 shows, for base stations, a hypothetical but typical arrangement of two frequency channels. Single-frequency channels would be coded in a similar manner, but would refer to the single frequency.

To identify when the reverse arrangement is used – i.e. when the transmitter frequency is used for receiving and vice versa – the prefix can for instance become a suffix if desired. Thus C21 becomes 21C. This normally affects only the allocation of frequencies of radio links, although, if a mobile is quoted out of context of a system, the method could be used.

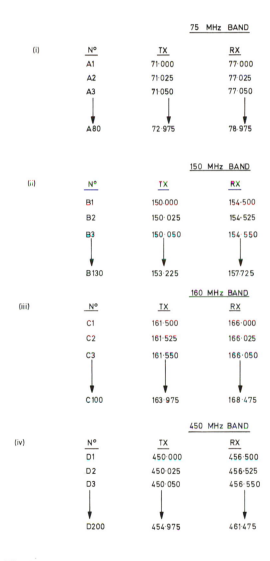

Fig. 21.3 Use of a Channel Numbering System

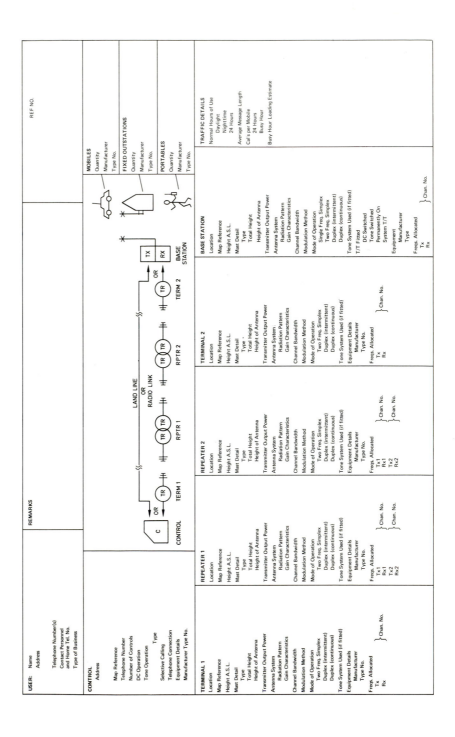

Fig. 21.4 System Block Diagram – Basic Form

21.13 Block diagrams of systems

Although, apparently, not relevant to record keeping, a simple block diagram of the overall system can be a great assistance in understanding the type of configuration involved and the distribution of the detail relative to the overall plan.

On the basis of the information which can be displayed on a single sheet, a suggested layout is shown in Fig. 21.4. In this arrangement, surplus detail has been included which in the simpler cases is deleted. Where a more complicated system is involved, either the system is broken down into several smaller arrangements, each based on the pattern of Fig. 21.4 or the customer is requested to supply a block diagram of his complete system which is attached to the simpler sheets containing the relevant detail.

The reference number in all cases is that of the application, suitably cross-referenced as in Section 21.11.

Fig. 21.5 to 21.7 shows a number of suitably marked examples using the format suggested above.

21.14 Computer techniques

The use of computers for keeping of frequency planning records is undoubtedly a logical development in the present technological age. Although of limited benefit in a small-scale allocation scheme, in a large country where the amount of stored information is considerable, these techniques come into their own.

Work is being undertaken by a number of frequency planning authorities (Ref. 2) to simplify the initial search procedure and at the same time to present to the operator alternatives which can be compared before a final choice is made. The final allocation can then be printed out in full for the main visual reference file.

Typical methods involve the insertion into the computer of the map references of the site specified in the application. Details of ranges from this site are then compared within the computer with other users in the vicinity and a number of channels – say 5 or 6 – are offered to the operator as satisfying necessary conditions such as third order intermodulation freedom and adjacent channel protection.

The operator now examines each of the offered channels together with the information on other users of that channel within the operational range obtainable from the site of the application. The business of each customer is displayed together with the distances. Also shown is the cumulative channel occupancy relative to various periods of the day.

On the basis of this information the operator chooses what he considers is the optimum channel and prints out the information for the visual file.

It should be pointed out however that these techniques are in the early stages of development and may change considerably over several more years. The process is extremely complex and requires all the basic parameters to be available to the computer program. Lack of essential information will greatly handicap such a tool.

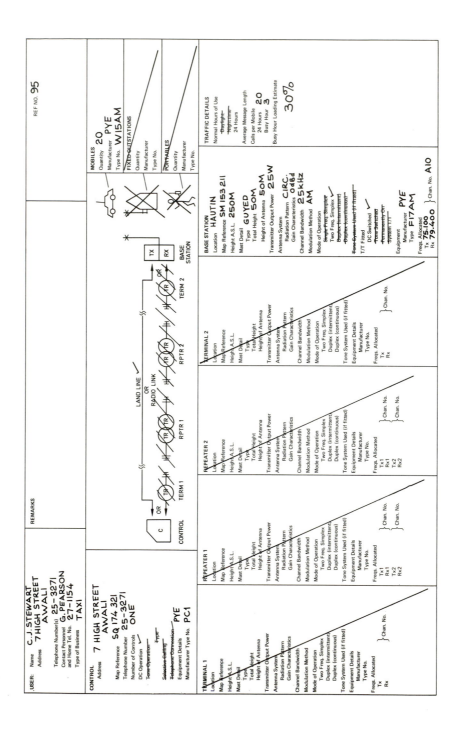

Fig. 21.5 System Block Diagram – Typical Land Line System

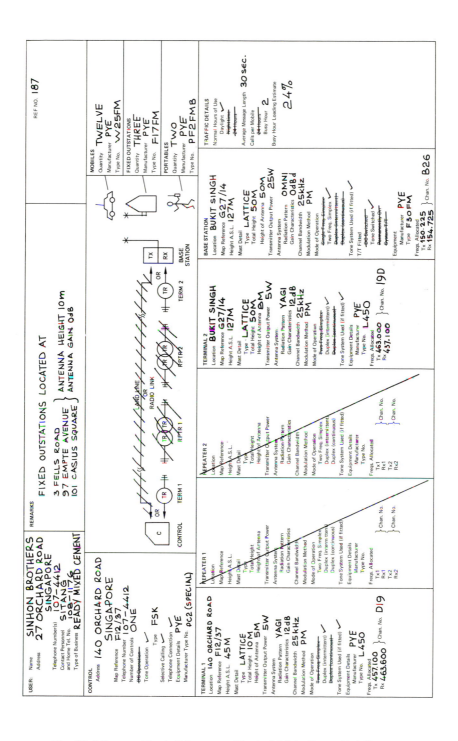

Fig. 21.6 System Block Diagram – Typical Link Controlled System

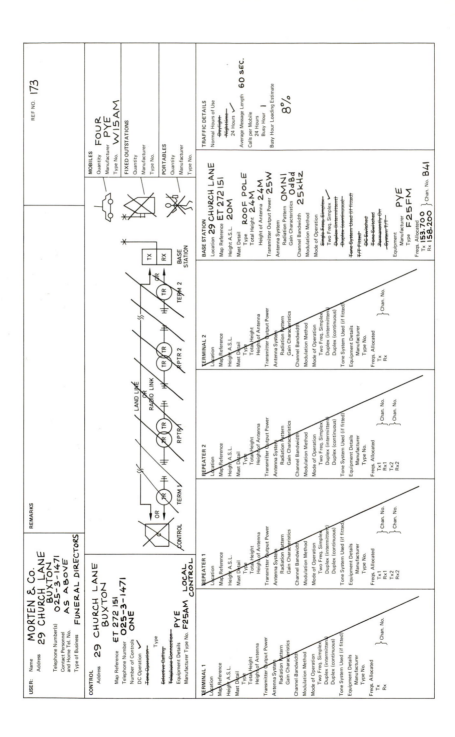

Fig. 21.7 System Block Diagram – Typical Local System

22. Monitoring of the Spectrum

There are two main reasons for monitoring radio channels:

1. "Policing" of the spectrum to check, amongst other things, illegal or unlicensed transmitters, obscene language, use of the channel outside the limits of licence and accuracy of the carrier frequency.
2. To obtain operating statistics such as types of message, message length, number of mobiles using each channel during the busy hour and times of busy hour.

Generally speaking the first category tends to be random; checks are made in various parts of the country at intervals determined by the facilities available. The equipment needs to be mobile and a suitably equipped van is often all that is necessary for a limited amount of spectrum monitoring.

The other category of monitoring is more sophisticated and could involve equipments in the centres of each of the main conurbations.

This guide does not aim to study these techniques; it is only intended to emphasise the need for some degree of monitoring to ensure that the spectrum is not misused but as far as possible is used in the most efficient manner.

23. Operators' Instruction Book

23.1 General requirements

In private mobile radio there are a number of operational hazards which can be easily avoided if all users are acquainted with the basic regulations and have a grounding in procedure from the outset.

Simple instructions are essential in many countries. Indeed, many manufacturers produce a booklet which, in well understood, non-technical language, amply illustrates the basic requirements.

23.2 Format

The booklet should be easily understood and pictorial instruction is strongly recommended. It should be small in size, concise and be relatively durable as it is liable to be kept amongst dirty and sharp objects on the dashboard shelf or door pocket.

23.3 Main Headings

These include:

(a) Radiotelephone system
(b) Mobile controls
(c) Portable controls
(d) Operating the various controls
(e) Calling procedure
(f) Message procedure
(g) What to do when out of communication
(h) Call signs and abbreviations
(i) Phonetic alphabet (English)
(j) Pronunciation of numbers
(k) Reporting of signal strength
(l) Local regulations
(m) Any other relevant information

Appendices

Appendix 1.   Shared Use of Mobile Radio Channels

1.   Introduction

Within a given area, effective usage of a channel is often maximised by combining a number of small users, each operating with relatively few mobiles or portables.

Time sharing of the channel in this manner raises a number of interesting points and the implications must be carefully considered for maximum use of the channel.

This appendix discusses these points and proposes means of optimisation.

2.   General requirements

The points deserving emphasis are:

(a)   Channel occupancy, involving:
   (i)   Busy hour distribution amongst users.
   (ii)   Types of message per user
      – average duration
      – importance of message relative to delay
      – distribution of messages during busy hour.
   (iii)   Average number of vehicles in use during busy hour per user.
   (iv)   Average number of messages per vehicle during busy hour per user.
   (v)   Competitors in same line of business.
   (vi)   Users with vastly different traffic requirements.
   (vii) User discipline.
(b)   Coverage requirement –
   (i)   Shared use versus single user.
   (ii)   Multiple users closely located.
   (iii)   Multiple users widely separated
      – effect of overlap on coverage pattern.
   (iv)   Single communal repeater.
(c)   Frequency considerations
   (i)   Netting of each user's equipment.
   (ii)   Netting between users.

(iii)   Effect of frequency differences on:
      – Channel in use
      – adjacent channels

3.   Channel Occupancy

3.1   The busy hour. The assessment of channel occupancy requires data obtained, initially from statistics applicable to the system itself or from averaged statistics obtained from similar types of systems. The former tends rather to be a post mortem and therefore the planning must be based on knowledge of the occupancy of similar existing systems suitable weighted to reflect operational differences.

Channel occupancy, is a broad and complex subject. References 10, 12, 14, 16 and 17 deal with aspects of channel loading. For the present, we will consider occupancy to be based on the traffic density during a fixed period in the busiest part of the day or night. In telephone practice, this period extends over one hour and occupancy is the fraction of that period the circuit concerned is loaded with traffic or held pending passage of traffic.

Mobile radio channels can equally be assessed by the same method although a shorter busy period span has often been advocated. In certain services, crowding of calls may occur at some point in the busy hour owing to extenuating circumstances. The rapid increase in calls on a taxi channel could coincide with, for example, a sudden change in weather, a sports event, or a theatre performance. They could all occur within a short period and thus modify the general pattern within the busy hour. However, for comparison between channels it is adequate in most cases to adhere to telephone practice and use the busy hour method as standard.

3.2   Single User v Multiple User. As shown in Appendix 5, the percentage channel occupancy is given as

$$\frac{t \; v \; c}{36} \; \%$$

Where t   =   average effective message length in seconds.
      v   =   average number of vehicles passing messages during the busy hour.
      c   =   average number of messages per vehicle during the busy hour (call rate).

Hence, given the necessary information, a reasonably close approximation of the busy hour loading can be obtained. Using the graphs, the busy hour loading can be related to Grade of Service, waiting to message length ratio, loss of message value and other factors.

If the information so obtained concerns a single user on an exclusive channel, then the busy hour period (or periods) can be identified. So the user can, if necessary and possible, re-arrange his traffic pattern or discipline to lower the peak loadings and reduce waiting times. This may even allow for additional mobiles.

However, in a shared circuit the position is quite different. The individual disciplines are unlikely to be related; indeed the effectiveness of each may be entirely different. Consequently the peaks could well be unnecessarily high owing to poor operating procedure. So a weighting factor will be applied to the total loading at times.

However one main advantage of shared systems contributes directly to spectrum economy. In exclusively single user systems, the busy hour peak is a function of the individual operation. Peaks involving a number of users on a shared system can, by suitable user choice, be arranged to provide a lower peak to average occupancy, allowing more mobiles to be accommodated on the channel before the peak hour loading is excessive. Fig. A1.1 shows such an arrangement.

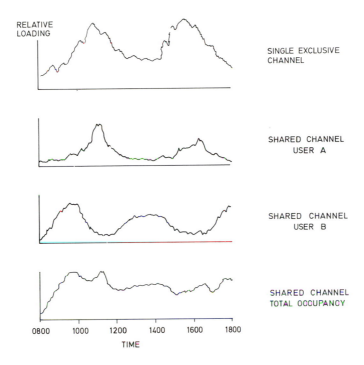

Fig. A1.1 Effect of Channel Sharing on Relative Loading in a Hypothetical Case

Thus we have a method, which can, notwithstanding inefficiency of individual users, by suitable grading, achieve a higher overall channel loading efficiency than normally obtained by a single exclusive channel.

Sharing of a channel by competitors in the same line of business may not be advisable or even practical. Therefore dissimilar users in the same area are often allocated the same channel on a time sharing basis, with, it is suggested, even greater advantage as the peaks are less likely to coincide.

4. Coverage requirements

4.1 Spatial distribution of multiple users. However, in spite of the gain in efficiency afforded by the above, there are factors which might offset the advantages so obtained.

In the case of the exclusive user, with a conventional system, the basic configuration may consist of a single site on which is located the base station transmitter and receiver. The area of coverage of this system will depend on antenna height and to a lesser extent transmitter power and antenna gain.

Let us now consider the shared system.

If the users are located within a short distance of one another, the total area

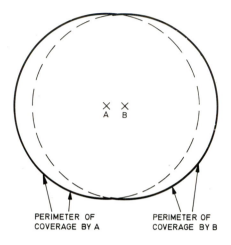

PERIMETER OF   PERIMETER OF
COVERAGE BY A   COVERAGE BY B

Fig. A1.2 Coverage of Two Closely Spaced Stations Sharing a Single Channel

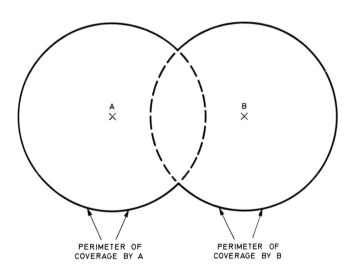

PERIMETER OF   PERIMETER OF
COVERAGE BY A   COVERAGE BY B

Fig. A1.3 Coverage of Two Transmitters Far Apart

covered by all stations would not greatly exceed that of the single user (see Fig. A1.2, where the total coverage perimeter is shown by the solid line). However, if the users are separated by an appreciable distance, then not only will the perimeter of the total area be the combined outline of the individual coverages (Figs. A1.3 and A1.4) but areas will exist in which mobiles will not hear the base transmitters of other users (Fig. A1.5) and message obliteration could result.

Thus we can postulate that users on a shared channel must remain within an acceptable distance of one another for the full advantages of sharing. The definition of "acceptable" is necessarily somewhat vague and, for equal coverage users, a maximum distance between any two users of 10% of the operational radius is suggested. Figs. A1.3 to A1.5 show how any appreciable separation between sites in

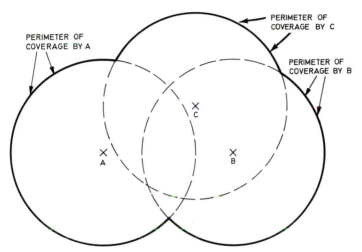

Fig. A1.4 Coverage of Three Transmitters Far Apart

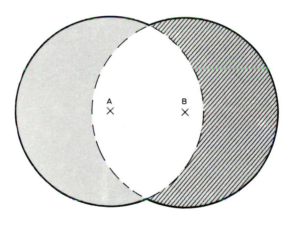

PROBABILITY OF LACK OF DISCIPLINE

AREA WITHIN COVERAGE OF A BUT OUTSIDE THAT OF B

AREA WITHIN COVERAGE OF B BUT OUTSIDE THAT OF A

UNSHADED AREA – COMMON TO A AND B.

Fig. A1.5 Effect of Shared User Site Separation on Discipline Probability a shared system can result in a marked increase in effective coverage area of that channel.

We take the simple formula showing range in kilometres from a base station to a mobile as:

$$4 \cdot 1 \left(\sqrt{h_1} + \sqrt{h_2} \right)$$

Where h_1 = base station antenna height in metres
h_2 = mobile antenna height in metres

Hence if the maximum deviation in radius is less than $\pm 5\%$ of the mean then it can be shown that the maximum increase, in total operational area, of the total number of sites involved need not exceed about 10%.

4.2   Effect on frequency re-use. As outlined in Appendix 2, for optional channel re-use, coverage should be minimised; otherwise, re-use points need excessive spacing which is wasteful of overall spectral resources. This can be shown quite clearly by examining Fig. A1.6. Considering a single station, the Appendix shows that, for sites having equal coverage, the re-use distance must be about 5 times the individual radius (Ref. 18).

Therefore a single site with an operational radius of 24 km should, assuming flat terrain, be separated from the next user of that channel by 120 km (see Fig. A1.6 (i)).

If, however, the coverage of a multiple user area is equivalent to an area of approximate radius 25 km *but* each user has a coverage radius of only 16 km, then the re-use distance in any direction need only be 80 km from the single user nearest to the re-use site. Fig. A1.6 (ii) illustrates this point.

Thus, although approximately the same total area is covered, the re-use factor is much improved.

4.3   Single Site, Multi-User configuration. Having shown many criteria for multiple site use in shared systems, an alternative should now be considered. This is the use of a single central equipment operated in a time-sharing mode by all users.

It pre-supposes that the users have been chosen for the best message/busy hour

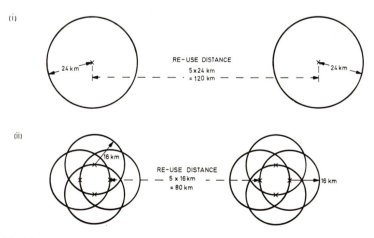

Fig. A1.6 Effect of Smaller Multiple Sites on Channel Re-use Distance

mix and that the required area coverages are similar and adequate for all users. If so, there is no basic reason why a communal equipment with suitable lock out and busy indication facilities cannot be employed. The method of control is however immaterial here. Some advantages of this method are:

(*a*)   A single operational radius rather than one determined by several separately sited systems is available.

(*b*)   As obliteration cannot occur with only one transmitter, some improvement in effective channel occupancy can result.

On the other hand, disadvantages can result, including:

(*a*)   If the users operating the system require appreciably different areas of coverage, then, to cater for all users, the antenna height must be increased. The resultant coverage area could approach that of a multiple site.

(*b*)   The single equipment, in a fully loaded system, will operate for a considerable portion of the busy hour. Therefore, if located on a communal site or near other units on nearby channels, it could produce a level of spectrum pollution high enough to introduce undesirable interference on other channels.

Where several such systems on nearby channels are co-located, the interference potential can be high.

(*c*)   As explained in 4.2 such an arrangement could require a much greater re-use distance than a number of single users individually covering portions of the total area.

5.   Frequency considerations

5.1   The cause of errors. Optimum use of the frequency spectrum depends not only upon suitable frequency allocations to users and their correct use but also upon the maintenance of these frequencies within prescribed tolerances.

With modern equipment design, accurate and consistent frequencies can be achieved.

However, means are usually provided within equipments to enable any small errors to be removed. This procedure, known as netting, ensures that all mobile receivers are tuned to the base transmitter and that all mobile transmitters agree with the frequency of the base receiver. It is usual for the base transmitter and receiver to be adjusted by means of a digital frequency counter, whilst the mobile units can be netted either by the same counter or by adjustment to zero beat using an IF crystal oscillator.

Either method is fairly simple and generally ensures a frequency deviation not more than a few Hertz from the nominal frequency. The process is normally repeated at intervals during the life of the equipment.

However in a shared system certain problems can arise:

(*a*)   New equipment introduced into a well established system will require frequency adjustment at rather more frequent intervals in the early stages than the other users. This is caused by crystal ageing which can be either positive, negative or a combination, according to the type of crystal. However, if the crystal obeys the conventional law of ageing, most of the frequency movement will occur during the first year and at 18 months or so the shift will be relatively small.

(*b*)   With a multiple user shared channel, the adjustment of all users to exactly the same frequency may be extremely difficult. One main reason is that users may employ different manufacturers' equipment which is, in turn, serviced by different organisations. So, the test equipment, such as frequency counters, could be different with consequent discrepancies. In other cases maintenance may be inadequate or non-existent, with servicing only when failure occurs.

5.2   The effect of errors. The first effect is that obliteration will be more noticeable.

As well as the usual effects of two or more mobiles radiating simultaneously, a marked frequency difference between carriers will distort still further the modulation of the users by superimposing beat notes. With AM there will be a constant heterodyne whistle, the frequency corresponding to the difference between the carriers, whilst with FM the beating is of a more complex form involving sideband effects.

As channel bandwidths decrease, the elimination of frequency errors becomes more important. At 12½ kHz the total bandwidth radiated (and received) must be closely controlled for maximum adjacent channel protection. Therefore any movement in the frequency of an equipment can cause it to affect or be affected by a signal on the adjacent channel in the immediate vicinity. The effect depends on the size of the errors. Fig. A1.7 illustrates the problem.

Summarising therefore, the probability of a multiple user shared channel causing or suffering from interference or obliteration is noticeably greater than in an

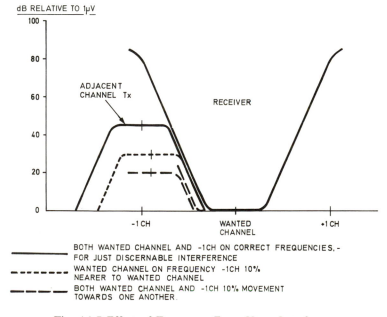

Fig. A1.7 Effect of Frequency Error Upon Interference

exclusive system. In the latter category we can however, include the communal single equipment for shared use.

6. Conclusions

This appendix shows how the shared user system can be optimised for channel occupancy, coverage and frequency effects. Channel occupancy and user distribution are important factors in shared systems. It is important to concentrate the users of a shared system within a small geographical area so as to minimise the total coverage. For most efficient re-use, the coverage area should be minimised. Accurate frequency control is essential in a shared system.

Appendix 2.   Assessing Frequency Re-use Criteria

1.   General

A number of reports have been issued by various authorities advocating extreme frequency conservation.

One report (Ref. 10) emphasises the need for the greater use of existing channels and indicates that the band available for mobile radio, although likely to increase around 1985, will be inadequate to provide discrete interference-free channels for every user. Reduced channel spacing may not be feasible. In an endeavour to accommodate future growth, mutual interference is likely between overlapping areas, particularly in the conurbations.

Another report (Ref. 17) points out the problems of overlap and its effect on the channel loading parameters. It emphasises the need for adequate channels to prevent such overlap effects and rejects the notion that overlap interference must be tolerated under certain conditions.

In view of the adverse effects of overlap and in the apparent absence of standard ways of assessing interference, this appendix sets out to consider ways in which overlap effects can be identified and suitable criteria defined.

Details given in a paper by Magnuski (Ref. 13) are used. A paper by the author (Ref. 18) should also be considered.

2.   Points highlighted by Magnuski

This paper is a recent contribution to the problems of spectrum conservation and shows, in its earlier sections, a method of calculating the spacing between sites employing the same channel. A limited degradation of 3 dB to any of the systems occupying the re-used areas is accepted in these calculations.

It shows an apparent reduction in distance between re-use systems employing FM compared with AM, but the example uses FM systems occupying 25 kHz channel bandwidth, and infers improved performance over the 12½ kHz systems currently in use in the U.K. It is postulated by the author of this Guide that a 12½ kHz FM system would however give very similar results and exhibit similar protection effects to those of 12½ kHz AM systems.

The paper by Magnuski also emphasises the capture effect of FM and, still referring to the 25 kHz system, states "It is conservatively assumed that an interfering signal 9 dB weaker will lower the S/N by less than 3 dB". With a 12½ kHz system this figure would be considerably in error, and would worsen considerably if fading were present (Ref. 18).

However the points raised in relation to AM appear relevant in the assessment of suitable re-use criteria and several of his conclusions are used.

Magnuski considers only a flat terrain throughout, and assumes that the signal strength falls off at a rate of 12 dB with each doubling of distance. He states that this is a compromise between the 6 dB free space loss and the inevitably higher rate as the horizon is approached. This assumption is agreed as a reasonable average and is therefore adopted in this appendix.

3.   Main points for consideration

3.1   Acceptable signal level. This has been conservatively taken as the level necessary to achieve 20 dB S/N under average conditions – with some local noise. On this basis a signal level at the receiver input of 3 μV is assumed.

3.2   Degradation by interference from re-use site. A figure of not more than 3 dB, as assumed by Magnuski, is adopted here.

3.3   Transmitter output power. This appendix considers only the fixed transmitter, assuming a figure of 20 watts. This is based on the U.K. specified maximum of 25 watts and an average feeder loss of 1 dB.

The mobile transmitter maximum output is taken as that which is sufficient for the type of system. It is assumed in all cases to be less than that of the fixed transmitter.

3.4   Antenna heights. Mobile. A height of 1·5 metres is assumed in all cases.

Fixed. This is a parameter determined in these notes, and is a function of site height, required coverage etc.

3.5   Frequency band. 150 MHz is taken as a basis for initial propagation calculations.

3.6   Terrain. As mentioned, Magnuski considers only plane earth. This appendix deals with site height as a parameter with corrections based on average terrain.

4.   Determination of Working range

A full propagation range study need not be made for every prospective site involved in frequency re-use. However, height approximation should be included to embrace the often encountered situation of high communal sites.

Fortunately with the parameters already defined, the simple formula giving range r_1 in kilometres as equal to $4\cdot1 (\sqrt{h_1} + \sqrt{h_2})$, where h_1 and h_2 are in metres, offers a close approximation to the actual coverage of an average system based on full path loss calculations and the parameters given in Para. 3. When a mobile antenna height (h_2) of 1·5 metres is assumed, we have –

$$r_1 \text{ (kms)} = 4\cdot1 \sqrt{h_1} + 5$$

where the range (r_1) corresponds to signal level into the mobile receiver of 3 μV.

The factor of 4·1 is based on average refraction (4/3) and is appropriate to the U.K. Where abnormal propagation conditions are frequent as in, say, sub-tropical zones, a different factor would be needed.

5.   Determination of unwanted signal level at the perimeter of re-use area

If we now use the 12 dB per octave of distance as suggested by Magnuski, the signal level will drop from 3 μV to –24 dB/3 μV at a range of four times that given in Para. 4. This implies an interfering signal at the perimeter of a re-use area of 0·18 μV. In practice, the actual figure will tend to be considerably lower unless the area is completely clear of local obstructions, in which case the primary signal at the perimeter of the re-use area will also tend to be higher than the required 3 μV.

As shown in Fig. A2.1, at the perimeter (effective or coverage range) of the re-use area A2, the interfering signal from the distant station A1 occupying the same channel should be less than 0·18 μV (a figure chosen so as to not degrade the 20 dB S/N ratio by more than 3 dB in a mobile receiver at the perimeter of A2).

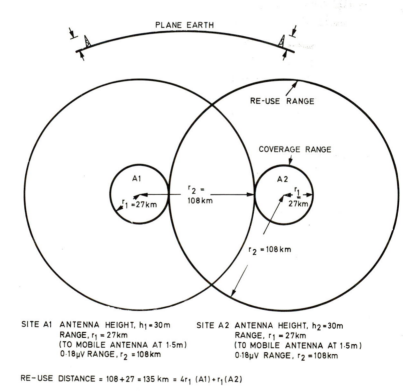

SITE A1 ANTENNA HEIGHT, h_1 = 30m
RANGE, r_1 = 27km
(TO MOBILE ANTENNA AT 1·5m)
0·18μV RANGE, r_2 = 108km

SITE A2 ANTENNA HEIGHT, h_2 = 30m
RANGE, r_1 = 27km
(TO MOBILE ANTENNA AT 1·5m)
0·18μV RANGE, r_2 = 108km

RE–USE DISTANCE = 108 + 27 = 135 km = $4r_1$ (A1) + r_1 (A2)

Fig. A2.1 Simple Two-Site Case (Assuming Plane Earth and Equal Antenna Heights)

6. Effect of dissimilar site coverage

Magnuski's treatment considers the operating range as merely a radius (r_1) around a fixed site, without a proper definition of other essential factors.

It is therefore necessary to introduce into the treatment such factors as antenna height, site height, terrain variations and different coverage requirements. Although, in the initial cases, an omnidirectional antenna pattern is assumed, directional arrays must also be considered.

Examining Fig. A.2.2 it can be seen that the use of a high site for A1 gives a much larger area than in the simple case previously considered. A reasonably flat surface is assumed for the intervening terrain.

The interference range limit must always be based on the station having the greater range. With two dissimilar sites, the one having the smaller operational range will necessarily give a smaller radius of interference well clear of the operational range perimeter of the other station.

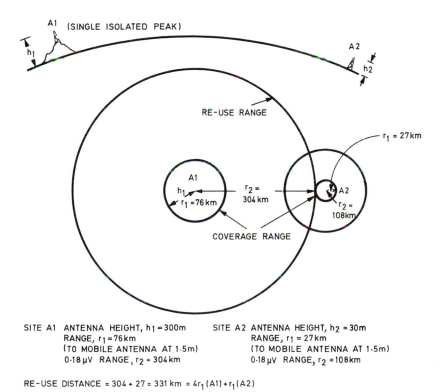

SITE A1 ANTENNA HEIGHT, h_1 = 300m
RANGE, r_1 = 76km
(TO MOBILE ANTENNA AT 1·5m)
0·18 μV RANGE, r_2 = 304km

SITE A2 ANTENNA HEIGHT, h_2 = 30m
RANGE, r_1 = 27km
(TO MOBILE ANTENNA AT 1·5m)
0·18 μV RANGE, r_2 = 108km

RE-USE DISTANCE = 304 + 27 = 331 km = $4r_1$ (A1) + r_1 (A2)

Fig. A2.2 Effect of High Site for One Transmitter

7. Effect of terrain

7.1 Undulating terrain. Assessing terrain is a complex procedure and to achieve a simple solution infers that a degree of averaging must be used. In Fig. A2.3, although A_1 is nominally at a height of h_A with A2 substantially at sea level, the intervening terrain introduces a modifying effect.

However, by using the same earth curve as for the plane earth condition, and allowing this curve to pass through a reasonable average of the terrain peaks, one is left with a resultant height at A1 (h_1) which is nearer the true *effective* height. Calculations of *operational* range using this height may be pessimistic as the actual height of A_1 tends to be nearer its local actual height for coverage purposes. Nevertheless, the interference radius (r_2) is affected by the series of peaks and presents what we really require, the distance before we can re-use the channel.

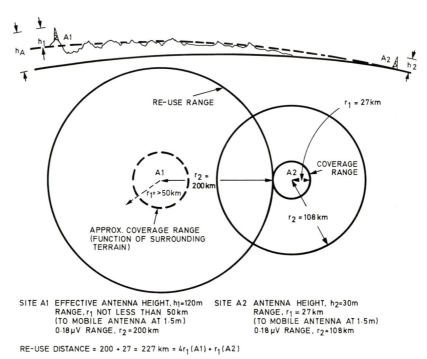

SITE A1 EFFECTIVE ANTENNA HEIGHT, h_1=120m   SITE A2 ANTENNA HEIGHT, h_2=30m
          RANGE, r_1 NOT LESS THAN 50km   RANGE, r_1 = 27km
          (TO MOBILE ANTENNA AT 1·5m)   (TO MOBILE ANTENNA AT 1·5m)
          0·18 μV RANGE, r_2 = 200km   0·18 μV RANGE, r_2 = 108km

RE-USE DISTANCE = 200 + 27 = 227 km = $4r_1$(A1) + r_1(A2)

Fig. A2.3 Terrain Adjusted Case

7.2 Mountainous terrain. In this situation natural obstacles can be utilised to provide a degree of isolation between sites using the same channel. However, effects such as reflections, ridge diffraction and atmospheric phenomena occasionally cause the reception of unwanted signals. Therefore reliance should not be placed solely upon a mountain range to isolate stations on the same channel, although it could well be that the mountain range, in any case, provides the distance protection described earlier.

Reference to Figs. A2.4 and A2.5 suggests that some allowance can be made in many cases but this allowance should not exceed 10 dB even with an apparently

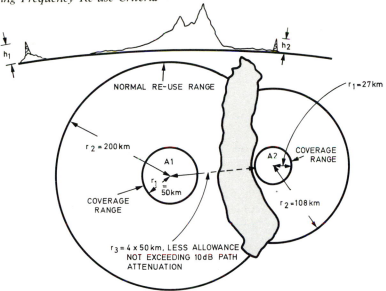

SITE A1 ANTENNA HEIGHT, h_1 = 120m
    RANGE, r_1 = 50km
    (TO MOBILE ANTENNA AT 1·5m)
    0·18µV RANGE, r_2 = 200km
    RANGE, $r_3 \triangleq 129$ km

SITE A2 ANTENNA HEIGHT, h_2 = 30m
    RANGE, r_1 = 27km
    (TO MOBILE ANTENNA AT 1·5m)
    0·18µV RANGE, r_2 =108km

RE-USE DISTANCE = 129 +27 = 156 km (FOR OBSTRUCTED PATH ONLY, OTHER PATHS 200+27=227km)

Fig. A2.4 Simple Obstructed Path

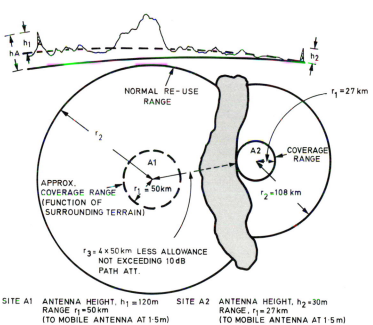

SITE A1 ANTENNA HEIGHT, h_1 = 120m
    RANGE r_1 = 50 km
    (TO MOBILE ANTENNA AT 1·5m)
    0·18µV RANGE, r_2 = 200km
    RANGE, $r_3 \triangleq 129$ km

SITE A2 ANTENNA HEIGHT, h_2 = 30m
    RANGE, r_1 = 27km
    (TO MOBILE ANTENNA AT 1·5m)
    0·18µV RANGE, r_2 = 108km

RE-USE DISTANCE = 129 +27 = 156 km (FOR OBSTRUCTED PATH ONLY, OTHER PATHS 200+27=227km)

Fig. A2.5 Obstructed Path With Terrain Adjustment

solid range of mountains splitting the path. The treatment can be based on plane earth or adjusted terrain according to which type predominates. Both are shown here; in the latter case, h_1 is corrected as a function of the terrain.

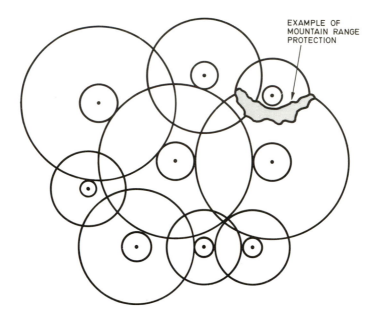

EXAMPLE OF
MOUNTAIN RANGE
PROTECTION

Fig. A2.6 Multiple Re-use

8. Multiple re-use

With sufficient area and observance of the procedures given previously, multiple re-use may be possible. The main point is that high sites dictate the degree of repetition. Therefore, sites should be chosen for the desired (but not excessive) coverage.

Fig. A2.6 shows multiple re-use techniques based on the principles given previously and includes an example of the protection afforded by a mountain range. All of the signal contours are free from overlap. The $0 \cdot 18 \, \mu V$ interference contour (outer circle) either coincides with or avoids the range perimeter (inner circle) according to the site position.

The coverage perimeter of some sites can be seen to coincide with the inter-ference limit boundary of *two* sources but it is considered that:

(a)   at the level chosen, interference from two or even three sources, is unlikely to be sufficient to blanket completely the wanted signal.

(b)   both interfering sites are extremely unlikely to radiate simultaneously on every occasion when the mobile at the perimeter of the wanted site nearest to the interference is receiving a signal from its own fixed transmitter.

(c)   all mobiles would have a squelch system adjusted to reject all signals below say, $0 \cdot 5 \, \mu V$, so that, during periods of listening-out, the distant fixed station(s) would not normally be heard.

9. Directional coverage patterns

Spectrum economy is generally improved by using directional antenna systems. This is particularly so where the unwanted direction involves area borders, sea boundaries, mountainous country, etc. In these cases, the general treatment is similar to that given previously. Fig. A2.7 shows an example of a multiple re-use arrangement with such a configuration. It will be noted that the directional interference boundary follows the pattern of the antenna. Fig. A2.8 shows the relationship between operational radius and antenna system gain (relative to a dipole). The actual range increase as the operational limit is approached will cut off more sharply if additional propagation losses accrue at the perimeter owing to vegetation or terrain.

To the rear and sides of the main lobe, the adjacent re-use sites are located in accordance with the principles previously discussed. However, with directional antennae, some care must be taken in assessing the distances in directions other than those covered by the main lobe. This is necessary because the exact pattern of the minor lobes cannot be easily determined. If in any doubt, the gain of an antenna in the doubtful directions should be considered as unity.

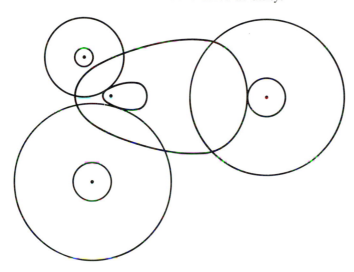

Fig. A2.7 Multiple Re-Use with a Directional Pattern

10. Effect of frequency band

As expected, the achieved operational and interference ranges are modified by the frequency band finally chosen.

The parameters discussed earlier are based on the 150 MHz band and require some slight adjustment if a materially different frequency is involved.

Considering the interference range, the fall off of signal relative to the operational range remains substantially as suggested by Magnuski (12 dB/octave of distance). On the other hand r_1, the operational radius, changes according to the frequency band used. Thus we have a simple method of adjusting the parameters to suit the frequency band.

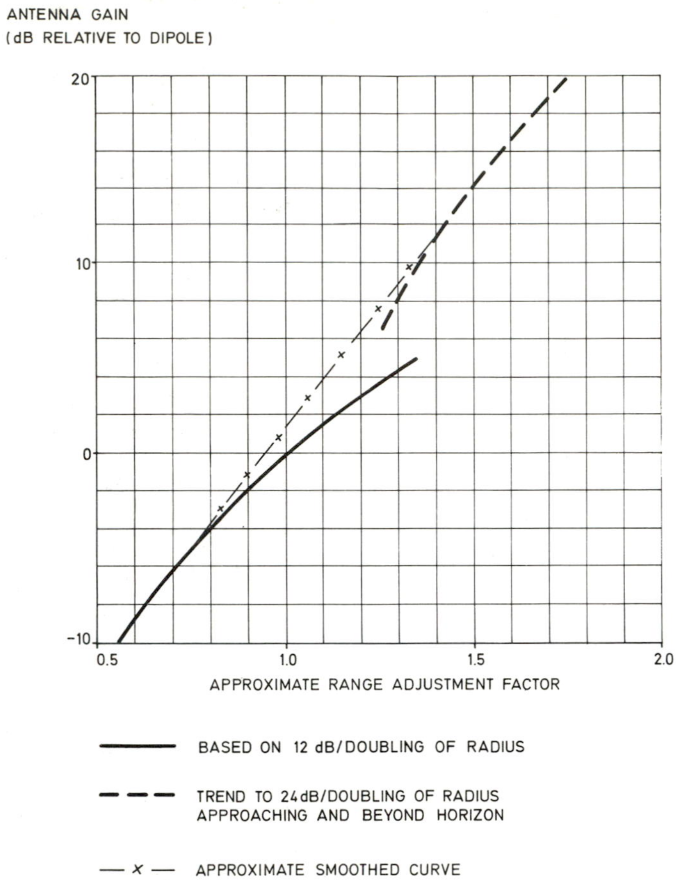

ANTENNA GAIN
(dB RELATIVE TO DIPOLE)

APPROXIMATE RANGE ADJUSTMENT FACTOR

—————— BASED ON 12 dB/DOUBLING OF RADIUS

— — — TREND TO 24 dB/DOUBLING OF RADIUS
        APPROACHING AND BEYOND HORIZON

— x — APPROXIMATE SMOOTHED CURVE

Fig. A2.8 Relationship Between Operational Radius and Antenna Gain

Assuming similar conditions, an approximation of the difference in path loss relative to 150 MHz (f_1) when the frequency band is changed to f_2 is given by

$$20 \log_{10} \frac{f_2}{f_1} \quad dB$$

For the purposes of this appendix, we examine two other bands, 75 and 450 MHz:

(i)   at 75 MHz, the approximate path loss will be 6 dB lower than at 150 MHz.
(ii) at 450 MHz, the approximate path loss will be 10 dB higher than at 150 MHz.

Translating these figures into actual range difference based on 12 dB per doubling of distance we have:

(iii) 75 MHz range factor 1·4 (increase)
(iv)   450 MHz range factor 0·56 (decrease)

These factors can now be included in the original simple formulae thus:

Operating range 75 MHz $5 \cdot 7 \sqrt{h_1} + 7$
in kms
150 MHz $4 \cdot 1 \sqrt{h_1} + 5$

450 MHz $2 \cdot 3 \sqrt{h_1} + 2 \cdot 8$

As stated earlier in this section, the interference range to the re-use site perimeter will remain equal to four times the above distance for any particular frequency band. Thus all calculations can be based on the operational ranges only.

11. Conclusions

The foregoing notes give a possible way of calculating the re-use distances of individual channels in the mobile radio frequency spectrum. They are based on certain well known propagation effects and include correction factors for height and terrain peculiarities as well as for different frequency bands.

   Whilst unusual propagation conditions involving the atmosphere or the terrain could inevitably affect the degree of interference, such eventualities have not been included and attempts to combat them would, most of the time, prejudice the advantages and simplicity of the proposed method.

Annexe A.   The Use of the Honeycomb Cell Method to Determine the Re-Use Distance Over Substantially Flat Terrain

1.   General

This annexe indicates a method of re-use distance calculation based on the cell method. The technique is becoming popular in systems designed for urban coverage in the higher UHF bands.

The method also has some merit in determining the re-use distance at lower frequencies and this annexe indicates the general solution based on such a technique over substantially flat terrain using limited coverage systems.

2.   Calculation of possible typical parameters

If we consider that a large area of reasonably flat terrain is to be occupied by a number of users requiring limited coverage and that the maximum operational radius of any one system is, for example, 15 km, then calculating back from the formula

Operational radius $= 4 \cdot 1 \left(\sqrt{h_1} + \sqrt{h_2} \right)$ km

where both heights h_1 and h_2 are in metres, we have, assuming a mobile effective height of one metre (h_2), a height h_1 of just over 7 metres. Alternatively, in Imperial measure we have:

Operational radius $= 1 \cdot 4 \left(\sqrt{h_1} + \sqrt{h_2} \right)$ miles

with h_1 and h_2 in feet.

(The constants are based on a frequency of 150 MHz.)

From this we derive an approximate radius of 10 miles with a fixed site *antenna* height of 25 feet.

We can now calculate that the re-use distance between sites over substantially flat terrain should be of the order of 75 km (or 50 miles).

3. Use of cell principle using above example

Examination of Fig. A2A.1 shows a hexagonal pattern based on a predetermined circle. The hexagons can be constructed by a compass with the chord length equal to the circle radius.

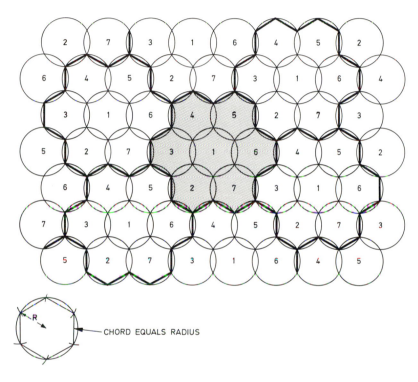

Fig. A2A.1 Seven Cell Frequency Distribution – Typical Section

If we now consider the central seven cells and, in each of these cells, allocate a separate channel, then by repeating the seven cell pattern in the manner shown, the channels can be seen to recur at regular intervals.

Measuring the spacing between the cell centres in which the same channel is repeated it will be seen that this spacing is about 4·6 times the cell radius – a figure very close to that of 5 times put forward as the re-use distance. If we artificially increase the radius of the drawn circle by just 9% (without actually increasing the system coverage) we obtain a distance between re-use areas equal to 5 times the actual operational coverage.

4. Application of cell technique

Following the principles mentioned above, a map of the area involved can be over-laid with a scaled transparent hexagonal grid to give a regular re-use pattern for the whole area. In fact several such grids may be independently used, each based on different operational radii. Thus channels for different services can be based in accordance to their actual coverage pattern.

In actual use, the cells can be applied, not only to individual channels, but also to frequency patterns (see Section 12 of Guide). The coverage pattern of a multiplicity of sites forming a single total area will, of course, be much greater than that of a single site and therefore the larger grid cells will imply greater re-use distances.

5. Other Configurations

A double row of protection cells is occasionally specified. This can be seen to consist of 19 cells giving a protection ratio of 7·5 times the operational radius of a single cell.

Appendix 3. The Effect of Unwanted Overlap Between Areas of Channel Re-Use

1. General

Channel congestion depends not only on the number of mobiles allocated but also on the degree of channel re-use.

This Appendix considers some consequences of excessive channel re-use, particularly in conurbations.

2. Co-channel interference relative to overlap

Consider first some overlap existing between two areas with a common channel. The effect on both the base station and the mobiles must therefore be assessed. It can be seen that overlap can cause different effects under different conditions.

With similar areas having 10% overlap, an average 10% of each area's mobiles occupying the channel might be located in the overlap at any moment. In practice it would depend on the types of service and the mobile distribution in each area. Thus the busy hour loading of *each* base station could be increased by up to 10% *non-coherent* loading. By non-coherent, we mean loading which is not under the control of the individual base station and where channel discipline is therefore unlikely to be followed.

Unfortunately this is not the end of the matter. In the overlap area, all mobiles would be subjected to the base transmissions from both *areas*. To the mobiles in the overlap this could be equivalent to a single base station loaded with twice the number of busy hour mobiles, half of which do not observe the same channel discipline.

Under these circumstances, there would be an extremely high lost call rate or waiting to holding time ratio. In fact, the area of overlap could be of virtually no operational use to mobiles of either net.

Thus it is imperative that overlap is minimised if considerable areas of the country (particularly within conurbations) are not to be rendered sterile to mobile users.

However, overlap cannot be completely eliminated in some terrains. A mobile located on a high ridge could well receive signals from more than one area. We do not imply that such a condition necessitates greater separation between conflicting sites. Indeed the high ridge may permit reduced physical separation in spite of a narrow overlap area near the top of the ridge.

3. Obliteration in overlapping Areas

Although discipline and caution are needed before transmitting on a shared common channel these precautions should not be needed for different areas. In other words a mobile in one area should not be prevented from operating normally by interference on his channel from adjacent areas. This, of course, may not apply during periods of abnormal propagation.

If obliteration is possible by the overlapping of areas, then a certain number of mobiles in an overlap may find it almost impossible to access their base station; the wanted base may also be obliterated for considerable periods by the adjacent area base station or stations.

4. Local Area Coverage

Overlap difficulties also cannot be ignored in local coverage UHF systems. Although the obvious propagation limitations of UHF do much to restrict the coverage of such systems, the ability of UHF multipath effects to penetrate into apparently screened areas must ultimately be influential in deciding workable re-use factors. The variable absorption of non-reinforced buildings with weather change must also be taken into account. City development programmes must also be considered in the planning of coverage.

5. Conclusions

Coverage overlap between channel re-use areas cannot be accepted as a general practice. Un-coordinated use by the two sets of users can bring about sterilisation of the overlap area and an increased loading, suggesting that it is an uneconomical planning decision.

Appendix 4.   How Much, How Far and for How Long.Can Radio Frequency Spectrum Usage be Assessed?

1.   General

There are three obvious dimensions which can be applied to radiation within the radio frequency spectrum. These are frequency, amplitude and time. In a sense, they are inter-related – frequency and time being the most obvious.

We are mainly concerned with their application to that part of the spectrum allocated for land mobile use.

This Appendix shows how these quantities may be defined.

2.   Normal channel parameters

A channel for use by a land mobile system can be described by the following:

- (a)   Its transmit and receive frequencies
- (b)   The mode of operation – AM, FM, PM etc.,
- (c)   The spacing between adjacent channels
- (d)   Modulation – speech, tone etc.,
- (e)   The transmitter powers allowed
- (f)   The antenna system gain, directivity, etc.,
- (g)   The fixed station height
- (h)   The number of mobiles to be used (channel occupancy)
- (i)   Type of message handled – short, long, etc.,

Based on these requirements, a frequency pair is allocated, isolated as far as possible from likely interference caused by other systems in the vicinity. The channel proposed may be shared by other users in the area and so the total channel occupancy may not be directly under the control of individual users. Also, in spite of frequency management precautions, interference may occur, both from nearby users on frequency related channels and from distant users occupying the same channel.

We will consider both the base station site and that of the individual mobile. Both have similar problems although frequency pollution is more likely from the base station in view of the greater antenna height. Pollution, caused by a mobile temporarily located on high ground, is relatively insignificant.

3.   The Frequency Dimension

This is possibly the most obvious of the three basic "dimensions", although not necessarily the easiest to control.

3.1   The transmitted carrier frequency from the base station. This is the nominal frequency on which the transmitter is permitted to radiate.

It is however, not a spot frequency. It is subject to movement, both long and short term. In the long term the nominal frequency will be determined by the stability of the source, affected by temperature and power supply changes, as well as long term settling effects (see Fig. A4.1 (i)).

In the short term, random fluctuations in current flow, particularly in the frequency generating source, will result in the carrier being modulated by noise components, thus effectively adding sidebands to an otherwise clear carrier (see Fig. A4.1 (ii)). Similarly, fluctuations in the power supply caused by inadequate smoothing, noise pulses, etc., can also add to the unwanted modulation level.

The sidebands attributable to noise extend well beyond the normal speech side-bands and have been quoted by Buesing (Ref. 5) as being around 45 dB, below the carrier at frequencies very close to the carrier, but improving to about –80 dB at ±15 kHz from the centre frequency.

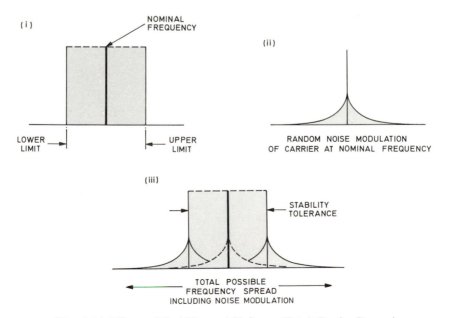

Fig. A4.1 Effects of Stability and Noise on Total Carrier Excursions

Thus, in spite of all normal precautions, we no longer have a single frequency radiation but one with noise modulation, albeit at a fairly low level, together with a long term movement (see Fig. A4.1 (iii)).

Unfortunately, this is not the end of the problem. Non-linearity in the transmitter, whether intentional (e.g. class C amplification) or unintentional, results in the generation of harmonics of the basic carrier frequency. These products will, although hopefully at a low level and over a short distance only, sterilise further parts of the frequency spectrum, thus effectively increasing the frequency dimensions (see Fig. A4.2 (i)). Antenna characteristics reduce even harmonics, but odd harmonics will often be radiated quite effectively.

A further effect of transmitter non-linearity occurs when several transmitters are located in close proximity. The coupling between transmitters (usually via the antennae) gives rise to intermodulation products. Still more frequencies are affected in the vicinity with a further increase in the frequency dimension. Fig. A4.2 (ii) illustrates effect.

These unwanted frequencies will also carry noise sidebands and be subject to the tolerances of the original frequency sources (see Fig. A4.2 (iii) and (iv)). Furthermore, where products are derived from harmonic relationships, both sidebands and long term movement of the carrier will be appropriately multiplied.

Reverting to the primary frequency, a further expansion in the use of the spectrum can occur when a single channel is occupied by a number of small users in the same general area. Unless each of these users transmits on *exactly* the same frequency, with long term drifts *exactly* coinciding, there will be a broadening of the effective frequency in that area with limits spanning the worst cases in each direction as shown in Fig. A4.3. Simultaneous radiation from more than one site (inevitable in spite of rigid discipline) can cause additional products together with

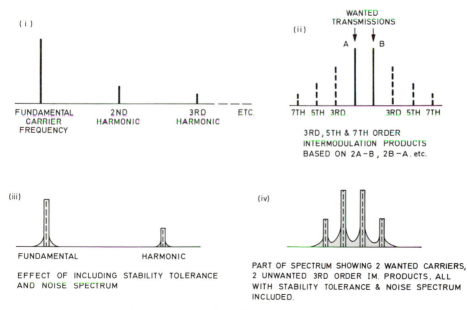

Fig. A4.2 Transmitter Spectral Components

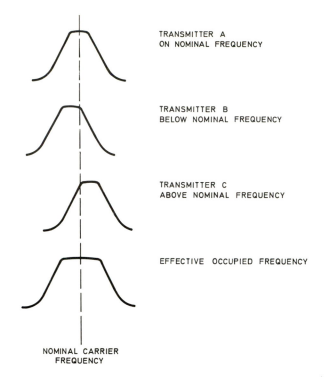

Fig. A4.3 Effective Frequency Occupation by Mis-tuned Transmitters

beat notes and additional noise. These effects will extend for a considerable distance, often embracing much of the combined areas.

3.2 The transmitted carrier from the mobile. All the effects described in 3.1 can also occur with the mobile transmitters. Frequencies are even less likely to coincide, particularly between mobiles allocated to different users, than with fixed transmitters. So the effective band occupied by the carrier (nominal carrier ± all tolerances ± transmitter noise) could well be considerably worse than with a single transmitter.

Fortunately the effective height of a mobile antenna is usually but a few feet and, therefore, large areas are far less likely to be affected. Only when, for instance, a number of mobiles are close to a communal site is excessive transmitter noise likely to be experienced by receivers on that site.

3.3 The modulated carrier frequency from the base station. With conventional land mobile equipment, modulation can take the form of speech or tone (on-off, FSK, PSK etc.,) or a combination of both.

3.3.1 With sinusoidal amplitude modulation the theoretical radiated bandwidth should not exceed twice the highest modulating frequency as shown in Fig. A4.4 (i). In practice, however, non-linearity can cause additional distortion and introduce harmonics in the source, the modulation and the modulated stage. Harmonics so generated will cause further sidebands to occur at the harmonic frequencies. The use of filters at strategic points in the chain can reduce these and at the same time

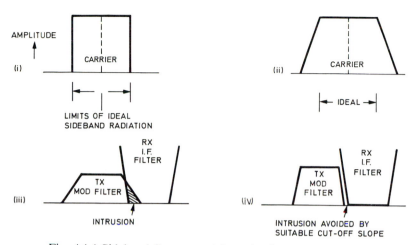

Fig. A4.4 Sideband Spectra and Intrusion into Receiver Passband

restrict the higher frequencies in the wanted waveform. The final result will never-theless include some additional unwanted sideband components depending upon the efficiency of the filtering devices. The sideband spectrum will therefore appear in the form shown in Fig. A4.4 (ii).

Fig. A4.4 (iii) shows that the pass band of such filtering should be adequate to permit the transmission of wanted audio components, whilst the cut-off slope should be similar to the receiver IF filter for optimum performance.

Fig. A4.5 shows how noise and frequency tolerance worsen the effects described above.

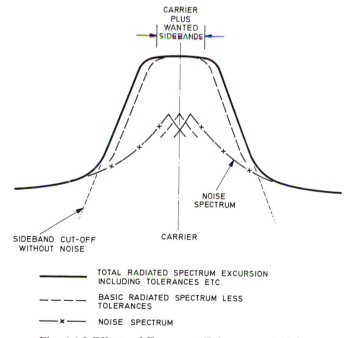

Fig. A4.5 Effects of Frequency Tolerance and Noise

3.3.2   Let us now look at the FM/PM case. The modulation index is Fd/Fm, where Fd is the deviation and Fm the modulation frequency, and determines the frequency excursion relative to the carrier.

A frequency modulated wave contains many frequency components.

The first component is analogous to the carrier frequency in an AM system but differs insofar as its amplitude is dependent upon the modulation index. The second pair of frequencies corresponds to the sidebands in an AM system but with the phase shifted by 90°. The amplitude of these components is not directly proportional to the modulating signal except at very small values of Fd/Fm.

The remaining components represent higher order sidebands that are not present in AM. When their amplitude is appreciable, the bandwidth has to be greater than AM *for the same intelligence*.

PM or phase modulation is similar to FM but here the modulation index is defined as the peak phase deviation of the carrier in radians. Thus, whilst with FM the modulation index varies inversely with frequency for a constant deviation, with PM the index depends only on the amplitude of the modulating signal and is independent of the modulating frequency. Thus, unless suitable precautions are taken, PM can exhibit a higher degree of sideband radiation than FM.

Summarising, unless a limited modulation index is employed together with suitable filtering, both systems can radiate a wider spectrum than an AM system carrying the same intelligence.

As with AM, the effective cut off of the sideband radiation should be of the form shown in Fig. A4.4(iv) whilst the effects of frequency stability will be as in Fig. A4.5.

Again, as in AM, the sideband spectrum of harmonics and intermodulation products will be determined by the characteristics of the primary sources whilst the long term stability will also influence any spurious products.

3.4   The modulated carrier frequency from the mobile. In general, the mobile transmitter will exhibit effects similar to those of the fixed unit. However, the differences described in 3.2 will also occur. Whilst many of the effects can occur under adverse conditions, the likelihood of these occurring is much less than with fixed systems.

3.5   Multiple transmissions. In both AM and FM, if a single channel is used by a number of nearby users, the spectrum is broadened, both by long term drift and beating between carriers and sidebands.

In this situation, the extremes must be considered, particularly in the vicinity of transmitters exhibiting these limits.

4.   The Amplitude Dimension

In terms of radiated energy, the amplitude dimension determines the range or area over which the frequency dimension discussed in 3. extends. It is obviously vital to ensure adequate but not excessive range with unwanted radiations extending over a minimum area.

Unfortunately, the amplitudes of unwanted products are, at best, usually proportional to the primary or wanted signal. In the extreme case they can worsen at a much greater rate as the wanted signal level increases.

Measures to reduce the level of these products are often therefore difficult or expensive, particularly in areas of high signal level.

Signal amplitude depends on a number of factors which are now discussed.

4.1 Transmitter power. The signal level at any fixed point depends directly on the transmitter power, provided of course all other parameters remain constant. Thus the transmitter power should be sufficient for the necessary level of signal at the range limit. The transmitted power should not greatly exceed the chosen level if the unwanted radiated products are to be minimised.

Although increase in power directly offsets path loss, the range is not inversely proportional to path loss. As the horizon is approached, the effect of transmitted power upon actual effective range is somewhat more complex.

Path loss over a range of a mile or two tends to show the normal free space propagation increase (about 6 dB per doubling of distance). Therefore within this distance, the range covered doubles each time the power is increased by a factor of 4. Beyond 1 to 2 miles, the figure increases to 12 dB per octave of distance requiring a 16 fold increase in transmitted power. This condition generally extends to the radio horizon although tending to increase more rapidly at the limit of range.

Beyond this limit, the path increment increases still further. A figure of 24 dB is postulated by Bullington (Ref. 6) and others as typical..

For most calculations the figure of 12 dB is a reasonable working average.

From the above it can be seen that, with a known level of the unwanted product, the area of spectrum pollution can be estimated. The above takes no account of path obstruction losses and other modifying influences. To include these would be impossible in the general case; they should however be borne in mind and may be included in particular applications if greater accuracy be required.

4.2 Antenna gain. Higher antenna gain is, in some respects, equivalent to increased transmitter power. It is used for a number of reasons. These include increasing the effective radiated power in the required direction(s) (at the expense of other areas), improving the effectiveness of a low power transmitter, compensating for feeder losses and increasing the signal level into a receiver.

Antenna gain can be influenced in two ways:

(*a*) Where the vertical gain ("E" plane in a vertically polarised antenna) is relatively unaffected, but the horizontal gain is adjusted to produce a directional pattern. Here the vertical distribution is not changed, the horizontal pattern merely being squeezed to form the desired shape.

(*b*) Where omnidirectional coverage is required, this is normally achieved at the expense of the vertical pattern. The "E" plane becomes thinner and the pattern is predominantly horizontal.

One advantage of increasing the *effective* transmitter output by antenna gain is the improvement in frequency discrimination. An antenna is designed to match its transmitter and feeder system over a narrow band of operating frequencies. At other frequencies, mis-matching occurs and efficiency will be lost; in effect the system is selective. This selectivity is often more pronounced with gain antennae. At *odd* harmonics of the design frequency, some degree of match will recur, often with a different radiation pattern.

A further advantage of using greater antenna gain is that, where unwanted radiated products tend to increase at a rate greater than that of the *actual* transmitter power, these unwanted products can be conversely reduced by using lower power transmitters and gain antennae.

4.3   Height of transmitter antenna. Antenna height is one of the most important criteria affecting the amplitude dimension. Here the term "effective height" refers to the antenna height *above average terrain* and therefore includes all or part of the actual site height.

Usually, the range of a transmitter in kilometres is approximately $4 \cdot 1 \, (\sqrt{h_1} + \sqrt{h_2})$ where h_1 and h_2 are the effective heights in metres of the base station and mobile. Thus it can be seen that, to the radio horizon, each octave of distance can be accomplished by increasing the effective height of the antenna by a factor of four. This assumes of course adequate transmitter power.

It is interesting to note that, at points well within the radio horizon, the effect of doubling antenna height is to reduce the path loss by 6 dB, and therefore the factor of four given above tends to confirm the figure of 12 dB per octave of distance at similar ranges (see 4.1). To define the above, we can say that, for antenna heights that are greater than the minimum effective height, the received signal is increased by approximately 6 dB every time the antenna height is doubled until free-space transmission is reached.

It is obvious that high sites, as well as radiating unwanted products over a wider area, can inhibit channel re-use. In short, this aspect of the amplitude dimension should be limited (see Appendix 2).

4.4   Frequency band. Although strictly within the dimension of frequency, the choice of a suitable band can influence the amplitude dimension.

Relative to that at 150 MHz, the loss at 75 MHz will be reduced by the order of 6 dB, whilst at 450 MHz the loss will increase by 10 dB. This is based on the formula

$$20 \, \log_{10} \frac{f_2}{f_1} \quad dB$$

where, in the example, f_2 equals 75 or 450 MHz and f_1 equals 150 MHz.

This change in path loss at 75 MHz is equivalent to an increase in range relative to 150 MHz of about $1 \cdot 4$ times. At 450 MHz the range would reduce by a factor of $1 \cdot 8$ times.

This reduction in range with increased frequency is of importance when one considers the interference range for the harmonics of the wanted frequency. In calculating channel re-use, the higher bands show obvious advantages of reduction in spacing assuming constant heights.

5.   The Time Dimension

Here is a dimension which, in theory, should be the easiest to measure; a transmitter is in either one of two states, on or off. But in practice this parameter is seldom measured. This is partly explained by the fact that practically all mobile channels use the simplex mode of operation and therefore a variable distribution of on/off periods exists.

Briefly, the total time a base station transmitter is radiating during a 24 hour period is the product of the average message length and the number of messages passed. In load calculations we are concerned with the *channel* occupancy (the *total* message period for both directions) but in this discussion we are mainly concerned with the fixed station transmitting and therefore an approximation of the average message length in this case would be (total message period)/2.

In many systems, the night traffic is relatively light or even non-existent, but, during the day, periods will exist when traffic peaks occur and the transmitter will therefore radiate for a greater total period of time. If during a peak period (or busy hour) a channel is *fully loaded*, then the base transmitter (assuming two frequency simplex) could conceivably be radiating for 30 minutes in the busy hour. (In the case of duplex or talkthrough repeater services, this figure would be 60 minutes).

Thus, during this period, the area around the base site would be subject to the wanted radiation, with part of the area close to the transmitter affected by the unwanted products.

If the channel is shared by several users in the same area, each user with a base station, then we have a number of overlapping service areas each with its own time dimension. In spite of attempts at discipline, there will be many periods during which multiple station spurious effects occur. We can approximate this to a single area having a radius equal to the extreme range achieved by all the users and a time dimension equal *at least* to the total of all the users during the busy hour. In practice, obliteration caused by more than one station transmitting could increase the time dimension considerably, resulting in, at the worst, a continuous effective transmission within the area and a corresponding high waiting time for channel access.

In practice, a *channel* loading factor of about 50% is considered commensurate with a reasonable waiting time before access and therefore (on the basis of equal message lengths in each direction) we conclude that a fixed transmitter could radiate for 25% of the busy hour.

Appendix 5 discusses loading in detail.

6.   The Received Signal

If we assume that the signals "transmitted" in the earlier sections of this paper are to be of use to the recipient, then the receiver must be:

   (a)   compatible with the signal source
   (b)   located within the system design area of the transmitter
   (c)   of adequate sensitivity to accept the signals available within the design area and to provide satisfactory output in terms of signal to noise etc.
   (d)   located where interference of all types is below the level for acceptable signal degradation.

The fixed receiver is less likely to suffer a varying level of interference than the mobile receiver, whose location is arbitrary.

For instance, the level of impulsive noise interference will be much higher in the mobile receiver, particularly in urban areas with high traffic densities. With care, a fixed antenna can be located out of the immediate high level area.

Additionally, because of its height, the fixed antenna will usually be far more distant from the impulsive noise source than in the mobile case with two vehicles side by side.

However this antenna height is a mixed blessing. Distant signals, preferably wanted but often unwanted, will be of higher level, whilst the cost of very high structures is often prohibitive. It is for the latter reason that high remote sites are often preferred, both to permit a lower antenna structure to be used and to avoid local interference.

Such sites are, unfortunately, relatively scarce. For suitable coverage of a city often only one or two sites can be used and therefore these sites are often occupied by a number of users. Each user, as well as using the portions of the spectrum allocated, radiates energy (theoretically at a very low level) on other frequencies. If these unwanted products coincide or provide the mechanism for the generation of other products to coincide with a wanted receive frequency, then that receiver will be adversely affected on that site.

Thus frequency management, bearing in mind the technical limitations of present day equipment, must ensure that co-sited stations are protected against such products.

For mobile receivers, such interference is, in the majority of cases, unpredictable. Only when located adjacent to a non-compatible site, is the problem likely to be severe and such occurrences are fortunately few.

The multi-user channel always causes the most troubles and message obliteration can be avoided only by discipline to minimise the time dimension.

Lastly there may be long distance interference caused by the re-use of the channel at too close a distance. Here air discipline is neither observed nor expected as the two users theoretically should be completely isolated (except possibly when a mobile is located on a high ridge).

Re-use of channels is discussed in Appendix 2 in which criteria are suggested to avoid overlap. A complete breakdown of communications can occur in the overlap area, particularly if the channel occupancy is high in either station.

7. Assessment of spectrum usage

Examination of the parameters discussed above, frequency, amplitude and time, suggests that spectrum usage might be quantified. The measures described earlier could be used as a basis of comparison between various systems and therefore as the basis of assessment of actual usage.

Several methods are indicated:

(a) Is it possible to derive a total "volume" expression using the three "dimensions"? For example, a direct product of frequency, amplitude and time.

In this approach, frequency could, for instance, be in the form of the ratio (radiated bandwidth)/(carrier frequency) and would be applied to the wanted and unwanted signals as required.

The amplitude parameter, involving transmitter power, antenna height, gain and directivity etc., would have to be referred to standard terms, e.g. 1 watt, isotropic or dipole antennae, etc., for proper comparison.

The time "dimension" could be directly given in a fraction of the total occupancy (Erlangs).

With this approach, we would see that the final expression would be directly proportional to certain terms, bandwidth, occupancy, antenna height, etc. and inversely proportional to the second power of the carrier frequency. The factor would be proportional to the square root of the individual values of quantities such as transmitter power and antenna gain.

The expression derived by this method would, however, be difficult to relate to different systems in other parts of the frequency spectrum as there is no true reference.

(b)   Relating the expression given in (a) to expressions derived from the use of "ideal" parameters.

The method will undoubtedly remove some of the main objections of (a) above but will, in the case of spectrum occupied by unwanted products, produce an expression tending to infinity and therefore comparison between different levels will not be possible.

(c)   Relating the expression given in (a) to expressions derived from systems meeting a defined specification limit.

This is a more practical method since it relates to what is possible rather than to what is ideally required.

The use of this method shows quite clearly the importance of limiting antenna height to that necessary for the particular system being considered, rather than allowing unlimited height to be used for all systems, regardless of range requirements.

Unwanted products are more easily compared by this method as they can be related to equipment specification limits rather than to the idealised limits of (b).

Other methods may be suggested by further examination of the requirements. For instance the chosen method may be changed according to the application. For example, unwanted products may be related to a different baseline in order to highlight certain aspects.

8.   Conclusions

The Appendix describes the various "dimensions" within which usage of the frequency spectrum is confined and shows how, by including them in a comparative calculation, spectrum usage may be assessed.

It describes the importance of the various parameters affecting these three basic "dimensions" and indicates the areas in which improvements can be made. The main source of spectrum pollution, the transmitter, is considered in detail. The effects of pollution upon the receiving equipment, and some of the advantages and disadvantages of high remote site operation are described.

Appendix 5.   Mobile Radio Channel Occupancy – its Assessment

1.   General

The value of a mobile radio channel is measured primarily by the savings ultimately achieved by its use.

Statistics have shown that radio circuit efficiency (influenced by channel occupancy) is prominent in determining the usefulness of a mobile radio system. Frustration, lost message value, wasted time and a poor grade of service are but a few of the results of excessive channel occupancy. This Appendix aims to show how the system criteria and occupancy are related and to indicate how they may be improved.

2.   Effective Channel Occupancy

Effective channel occupancy is based on the level of traffic density existing during the busy hour.

The effective channel occupancy is derived from the total fraction of the busy hour during which the channel is loaded with traffic or held pending passage of traffic (i.e. including normal transmit to receive changeover periods).

In order to ascertain the traffic density during the busy hour it is necessary to define both the average length of an individual call or message period and the average number of calls handled during that time. The latter will be the product of the average number of vehicles involved and the average number of messages from each mobile. Thus the occupancy during the busy hour can be shown as

$$\frac{t\,v\,c}{36}\ \%$$

Where t = average effective message length in seconds.

v = average number of vehicles passing messages during the busy hour.

c = average number of messages per vehicle (call rate).

3. Message Length

Two parameters must be considered:

(*a*)   *Actual message length*
This is the total time required to transmit the actual information content of the message.

(*b*)   *Effective message length*
This includes (*a*) above together with call establishment, operational breaks and changeover periods.

In both cases the term "message" refers to the complete two way transmission. In the case of effective message length this extends from the beginning of the call to the final acknowledgment.

3.1   Average effective message length. This is the average total time the channel is occupied whilst passing a single message. It includes the periods necessary to establish the call, to pass the message and to terminate the call. It also includes the time necessary to account for poor operating, information retrieval and change-over periods.

Therefore channel occupancy is based on this time.

The average effective message length is strongly dependant upon operating efficiency, the ease of information retrieval and speech hesitancy, (the "umm's" and "er's" associated with inadequate grasp of the message content and the inability of the operator to control the net increase the effective message length).

Message repetition also wastes channel occupancy. Whether this is due to mis-understanding or poor reception is immaterial. Usually the first is caused by poor operating, whilst the second can be attributed to many factors, including inadequate coverage, excessive noise (electrical or acoustical) and faulty equipment.

The average effective message length is usually derived from statistics taken over a period of time involving a number of different users. More than one grade of message length may be desirable when optimising channel occupancy.

For instance an average can be derived from the statistics of all types of message and used for calculating channel occupancy for the mobile radio spectrum as a whole.

Alternatively, from two or more grades of effective message length, occupancy figures can be used selectively amongst the different channels in an endeavour to obtain high efficiency per channel. For example, effective message lengths appear to vary between 10 seconds and 60 seconds and may sometimes be even greater. An overall average of around 30 seconds therefore exists. However, a single user occupying a channel and passing short messages can obviously handle more vehicles than the user with the longer message content or the shared channel with several grades of users.

Hence, for maximum efficiency in shared systems, the average effective message length on which the occupancy of a channel is calculated must be based on the use of the smallest possible spread of individual effective message lengths.

3.2   Average actual message length. The time occupied by the total useful intelligence of a message can be termed the actual message length. Obviously it is beneficial to keep this period as short as possible. Measures involving abbreviations, code references and, latterly, data methods are all steps in the right direction.

It is therefore essential that the actual message length forms the major part of the effective message length, i.e. the ratio (effective message length)/(actual message length) should be near to unity as possible. Improved operating can reduce both the actual message length and the "setting up" period forming the difference between effective and actual message lengths and should therefore be considered very carefully in the effort to increase efficiency.

The average actual message length is obtained from measurements of a variety of users.

4. Busy hour vehicles

To calculate channel occupancy, the average number of vehicles using the channel during the busy hour is required. This is extremely difficult to define; much depends upon the operational mode, weather, environment and type of business. Often the smaller user has a greater proportion of his mobiles in use during the busy hour than the larger operator and the average vehicle number can vary by as much as 10 to 1. However results show that the average number of mobiles calling during the busy hour could be 25 to 30% of the licensed quantity.

Care should be taken to eliminate from any calculation obvious non-typical cases; on the other hand their appearance in any statistics should be regarded with some caution and further checks made to ascertain their validity before rejecting. It may well be, for instance, that a very low busy hour usage of a channel by an organisation having a high number of licensed mobiles is a seasonal effect and that a different time of the year could yield the opposite result.

Unless a statistic is the result of continuous monitoring and analysis, such errors as above are always likely. Even if the statistics are apparently complete, care must be taken in interpreting the results.

Work undertaken by Plotkin (Ref. 16), using a mathematical model, indicates that with average message lengths, the busy hour mobile complement plotted against channel occupancy is of the form shown in Fig. A5.1. With the total average effective length of message employed in this case, the number of users is approximately 55 per Erlang of traffic (100% occupancy). This is a little lower than the 65 users per Erlang of traffic postulated by McClure (Ref. 14) and the 63 observed in the mobile telephone service in Washington D.C. and referred to by McClure in his paper. This indicates that the Plotkin estimate of about 30 seconds for average message length may be slightly too high and a figure of 25 seconds may be nearer the true average. Alternatively, the average call rate assumed by Plotkin approximating to 2·18 per mobile is slightly too high and should be nearer to 1·85 per mobile. On the other hand, the Home Office Report (Ref. 10) indicates that, taken over a total of 17 users in the London area, the average Effective Holding time (equivalent to the average effective message length in these notes) is 34·7 seconds. The average call rate per mobile is also shown at the higher figure of 2·37. This implies 44 users per Erlang of traffic. Fig. A5.1 compares these results.

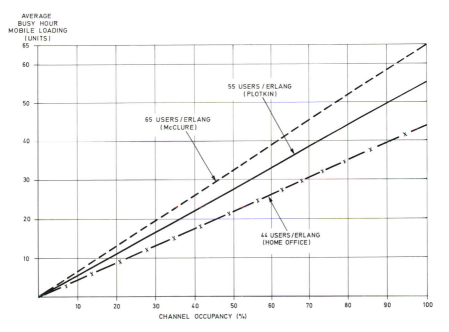

Fig. A5.1 Busy Hour Complement as a Function of Channel Occupancy

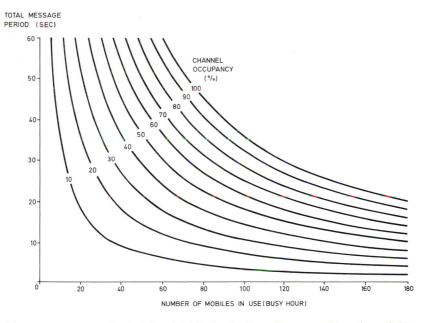

Fig. A5.2 Total Message Period Per Mobile in the Busy Hour as a Function of Channel Occupancy and Number of Mobiles in Use

5. Average Call Rate

Within the busy hour the total occupancy is determined by the number and length of the messages passed. During that period, if 'n' mobiles are operating, the average call rate per mobile will be (number of messages)/n.

Thus figures depend upon the type of business and the traffic passed. Taking the example given by taxi organisations (particularly in city areas), the number of messages from individual vehicles could well be high owing to the large number of short journeys made. On the other hand, the individual messages will also be short. In contrast, an electrical servicing organisation may pass long lists of address details to individual mobiles and this may happen only once or twice during the whole working period. Obviously there is a much lower average call rate in this case.

Fig. A5.2 shows the *total* message period or the sum of individual effective message lengths which can be allocated to each mobile, plotted against the quantity of mobiles using the channel during the busy hour. Values of channel occupancy from 10 to 100% are shown. Dividing the time of the *total* effective message period by the average effective message length will indicate the call rate per mobile. Take for example 40 mobiles operating during the busy hour with a calculated channel occupancy of 40%. The total message period will be 36 seconds. If an average effective 15 second message is assumed, this means that each mobile will call, during the busy hour, an average of 36/15 or 2·4 times.

6. Effects of excessive channel occupancy

From the formula for the percentage occupancy (t v c/36), it can be seen that there is a direct relation to traffic flow in telephone circuits as expressed by Erlang (Ref. 11).

Erlang has derived formulae for the calculation of traffic flow relative to the number of trunks required in telephone systems and the basic unit (an Erlang) has been accepted as a measure of the continuous traffic flow in a circuit for a given period, usually taken at the peak period or busy hour. Thus 100% occupancy of a circuit corresponds to one Erlang.

By reference to Erlang's formula or preferably by using the various charts summarising the formula, one can derive a grade of service factor from which can be obtained the average number of attempts needed to obtain access to a channel.

If access cannot be obtained immediately then a waiting period is inferred before any message can be passed.

This in turn introduces three adverse effects:

(*a*)   Time is wasted in waiting for channel access.
(*b*)   The impact or value of the message to be passed is usually lost as waiting time increases.
(*c*)   Increased obliteration often results from lack of access since several mobiles may attempt to gain access immediately a previous call is completed. This results in further wasted time whilst the base station controller "unscrambles" the calls and identifies and advises the next user.

6.1   Waiting time to average effective message time. In mobile radiotelephone systems, rather than hanging up and trying again later, the mobile usually continues to listen out until the mobile using the channel completes his or her call. An attempt is then made to gain access to the circuit.

Systems using this procedure consider the waiting time to average effective message length ratio to be the occupancy criterion rather than deriving it from the possibility of obtaining access to a circuit at random. The ratio of waiting time (W) to average effective message length (M_e) (Fig. A5.3) follows an exponential curve $\dfrac{W}{M_e} = \dfrac{p}{1-p}$, based on channel occupancy in Erlangs (p). It can be appreciated that the waiting time for a fixed ratio depends upon the message content, with long messages producing correspondingly long waiting periods.

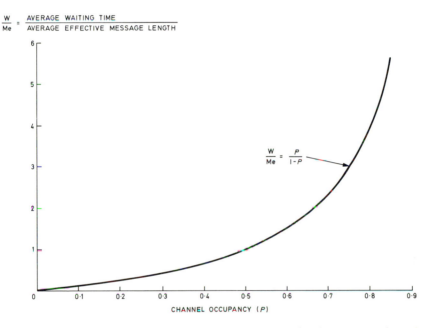

Fig. A5.3 The Ratio of Average Waiting Time to Average Effective Message Length as a Function of Channel Occupancy

It is thus in the interests of all users that, to avoid unbalanced utilisation, individual channels should be occupied by users passing messages of similar length.

Fig. A5.4 shows the average message delay as a function of channel occupancy. Three examples of average effective message length have been illustrated.

6.2   Message Value. The loss of message value is often underestimated by many users. As radio communication is intended to improve efficiency within a business organisation, it follows that delays in the passing of information must detract from the efficiency so gained. The degree by which the delay affects the message value is determined by the kind of operation and in taxi organisations, for instance, a message or instruction rapidly loses its value if delay occurs before it can be passed. On the other hand, messages in other business categories often have longer effective life and the value lost is correspondingly lower.

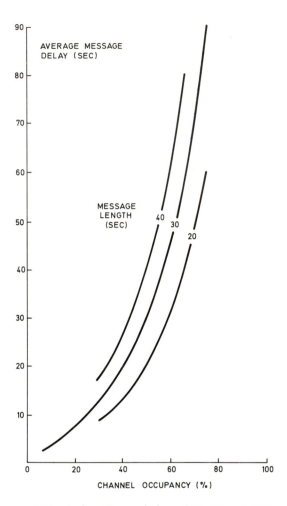

Fig. A5.4 Average Delay before Transmission of 20, 30 and 40 Second Messages

Plotkin (Ref. 16) has derived a formula, based on a mathematical model, showing the average lost value to be an exponential function of message delay or waiting time (t, mins). Fig. A5.5 shows curves based on this formula, i.e.:

$$\text{Loss of message value} = 100 \times \left(1 - e^{\frac{-t}{20}} \right) \%$$

Fig. A5.6 gives curves showing the loss of message value plotted against channel occupancy for three values of average effective message length, 20, 30 and 40 seconds.

However, although the loss in message value has been derived from the mathematical model of Plotkin and centres around an average effective message length of about 30 seconds, small deviations from this message length have, in fact, been assumed in some of the figures.

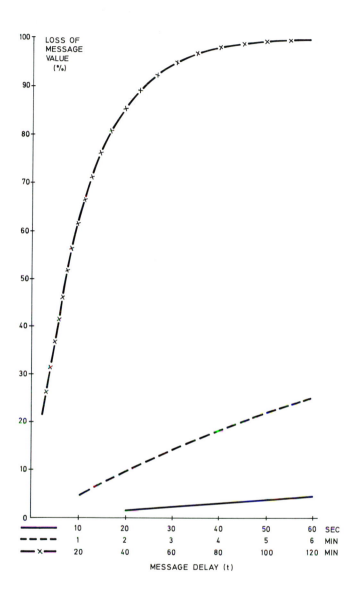

Fig. A5.5 Loss of Message Value as a Function of Transmission Delay

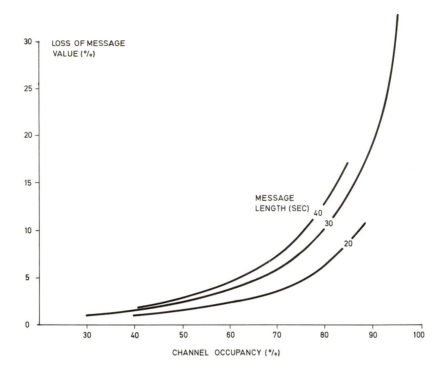

Fig. A5.6 Loss of Message Value as a Function of Channel Occupancy

6.3   Message obliteration. Where a number of mobiles operate on a common channel, the possibility of more than one mobile transmitting at one time cannot be ignored. In systems with tight operator control, this is much less likely during the passing of a message than in, say, a shared system with a number of smaller, separately controlled nets.

The amount of simultaneous transmission, regardless of the degree of control, must depend on occupancy. The most likely period when this can cause obliteration is directly after a period when control has been operating to a particular mobile.

This is the time when mobiles may be waiting to gain access to the net. When the channel occupancy is high, more than one mobile may be waiting so there will be an effort by all to obtain access immediately the original call is completed. To unravel this, requires both effort and time on the part of the base station operator. So there is more channel congestion than that on which the channel occupancy was originally planned. With shared systems where the individual sub-systems occupy partially overlapping areas, obliteration is even more likely.

The assessment of average effective message length must therefore include such periods as are necessary to clear the obliteration. This must inevitably worsen the ratio given in 3.2 and further increase the average delay before the message can be transmitted. As a result loss in message value will increase.

The above effect concerns all types of system under normal operating conditions. However, with FM systems an additional effect tends to nullify some of the alleged advantages of capture in providing a clear interference free signal (see Annexe A).

When a number of mobiles make simultaneous calls to establish a message sequence, the mobile nearest to the base station will capture or tend to capture, according to the relative signal levels. The result is that the more distant stations may not be heard at all, giving a further delay in establishing their call.

The JTAC (Joint Technical Advisory Committee) Report (Ref. 12) touches on this subject and two figures in this reference give typical examples of the effect. Each figure shows two sets of curves, one taken from recordings made of a typically loaded channel; the other set is based on a formula derived from a typical message format. The difference between the two sets of curves has been ascribed to the deviation of the recorded message from the mathematical model.

7. Grade of Service

Grade of service as applied to telephone networks is a measure of the ability to obtain access to the desired subscriber. It is usually expressed in decimal form, taking the reciprocal so as to indicate the likelihood of a lost call. For example a grade of service equal to 0·1 will also indicate one lost call occurring in ten random attempts to establish a call. The description "lost call" refers to a call which cannot be placed owing to part of the circuit path being occupied.

Where a radio telephone is connected into a land telephone system there is no reason to alter this form of measurement. Furthermore, its use in conventional mobile radio systems would seem to have some merit.

Standard telephone grading curves are based on formulae by Erlang and in the simplest form (for a single trunk or channel) the grade of service is given as

$$\frac{T}{1 + T}$$

where T is the traffic flow in Erlangs.

Alternatively, the lost call rate can be expressed as equivalent to one lost call in $(1 + T)/T$ random attempts.

In many of the figures accompanying these notes, channel occupancy is expressed as a percentage. As mentioned earlier, however, 100% busy hour occupancy is equivalent to one Erlang for a single channel. One Erlang is the traffic flow in a circuit or channel continuously occupied for a given period of time (busy hour). By expressing fractions of full occupancy in Erlangs, a direct relation therefore exists between channel occupancy and grade of service. Fig. A5.7 compares these quantities. For example, with an occupancy of 40%, the corresponding grade of service is 0·28. Since grade of service is the reciprocal of the likelihood of the channel being occupied then, in this case, for one in 3·6 occasions (at random), the channel will be occupied (i.e. the call will be lost).

Fig. A5.3 shows the variations of average waiting time to average effective message length (or average circuit holding time) plotted against channel occupancy. It can be seen that this is another way of expressing the effectiveness of a circuit. Next, Fig. A5.8 shows the effect of channel occupancy on lost call rate and it is interesting finally to examine Fig. A5.9, which shows the combination of Figs. A5.3 and A5.8 and expresses the ratio (average waiting time (W))/(average effective message length (M_e)) in terms of

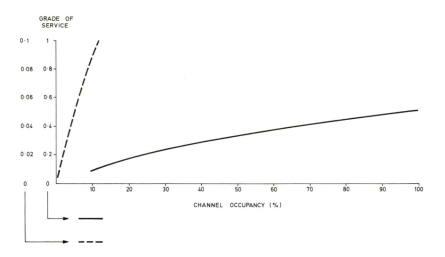

Fig. A5.7 Grade of Service as a Function of Channel Occupancy

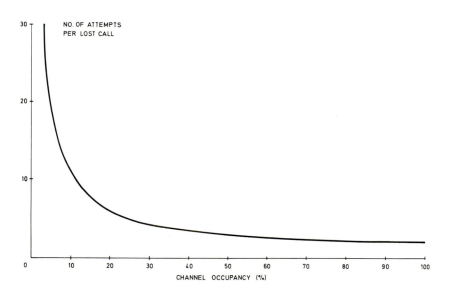

Fig. A5.8 Lost Call Rate as a Function of Channel Occupancy

lost call rate. It can be seen that this ratio starts to increase very rapidly when the lost call rate is worse than about 1 in 3 or 4 random attempts (approximately equal to a channel occupancy of 40%).

8.   Effect of Using more than one Channel

When one channel has become loaded to the maximum acceptable level, the possibility of using additional channels can be considered. It is not intended to consider here all the variations. However, it is sufficient to say that, with further channels, additional mobiles can be handled. Whether the additional number is directly or increasing proportional to the extra channels depends upon the method adopted. The following paragraphs discuss the latter possibility.

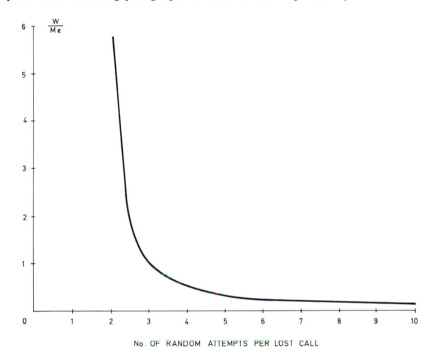

No. OF RANDOM ATTEMPTS PER LOST CALL

Fig. A5.9 Relation Between $\dfrac{W}{Me}$ and Number of Random Attempts per Lost Call

8.1 Basic method. If a mobile is equipped to operate on a number of channels and able to receive a call on any one of the channels at will, then, with certain system provisions, the ability to contact or receive on any unoccupied channel naturally improves the system efficiency.

8.2 Grade of Service. The improvement can be understood by reference to 7 (grade of service). If the possibility of a lost call with one channel is say, 1 in 4, then, with random access using 4 times the number of mobiles but employing four channels, the grade of service will improve to 1 lost call in 12·5. By considering simply the probability of all four channels being occupied simultaneously, the improvement can be visualised. Alternatively, if the original grade of service for one channel is considered adequate, then by random access to four channels as in the example, the total number of mobiles over four channels can be increased by a factor of approximately 10 times relative to one channel, four times due to the direct increase in the number of channels and 2·5 times brought about by random access, or trunking as it is often called. Fig. A5.10 shows the total traffic which can be carried over a total of from one to four channels or trunks compared with the traffic over each individual trunk. Fig. A5.11 shows the grade of service over one to four trunks; the numerical example given above can be easily verified from these graphs.

It should be noted that the information is derived from telephone practice which has similar problems of loading, trunking etc.

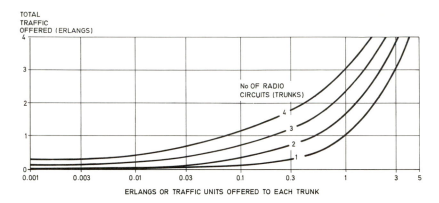

Fig. A5.10 Total Traffic Offered as a Function of Traffic to Each Trunk

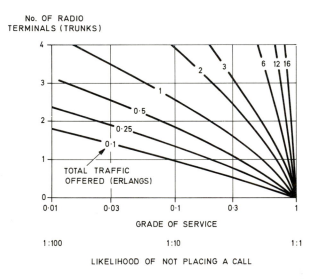

Fig. A5.11 Total Traffic Offered as a Function of Grade of Service

8.3   Calculation of Channel loading over trunked circuits. As in the single channel case it is essential to know the individual call duration (average effective message length) and the average call rate per mobile during the busy hour. Both must be derived initially from measurements.

To obtain these details, the average effective message length and the average number of calls per mobile over a 24 hour period can first of all be calculated from the records. The product of these two indicates the total time the channel is occupied during the 24 hour period by each mobile.

However, the distribution throughout the 24 hours will not be uniform, so the next requirement is to introduce a factor based on peak hour occupancy. In other

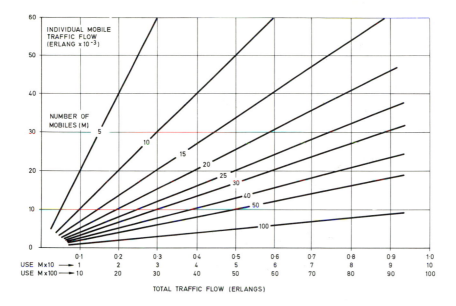

Fig. A5.12 Total Traffic Flow as a Function of Individual Traffic Flow for Various Quantities of Mobiles

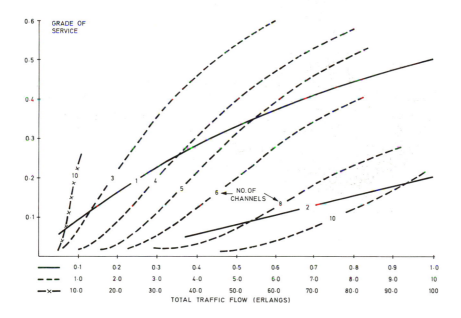

Fig. A5.13 Grade of Service as a Function of Traffic Flow for Different Channel Quantities

words, to ascertain the proportion of total calls during the 24 hour period which will fall in the busy hour period.

Alternatively the busy hour period is identified and analysed and the average call rate derived directly from the statistics. Whichever method is adopted, however, is immaterial to the final solution.

Let us at this point take an example to illustrate the total procedure.

(a)   Average effective message length = 30 seconds
(b)   Number of messages during 24 hour period = 9
(c)   Percentage of calls (b) in busy hour = 20%
(d)   Average busy hour call rate = $0 \cdot 2 \times 9 = 1 \cdot 8$
(e)   Traffic load per mobile in the busy hour = $(30 \times 1 \cdot 8)/3600$ Erlangs
       = $0 \cdot 015$ Erlangs

On the basis of this traffic level, the loading will be $1/0 \cdot 015$ mobiles per Erlang, or $66 \cdot 6$ mobiles/Erlang, which is the number of mobiles which will provide 100% channel occupancy during the busy hour. Thus with 40% occupancy (equal to $0 \cdot 4$ Erlang) the busy hour mobile loading will be $0 \cdot 4 \times 66 \cdot 6 = 26 \cdot 6$ mobiles.

Continuing, with this level of loading, the grade of service will be $0 \cdot 28$ (Fig. A5.7) giving a lost call rate of 1 in $3 \cdot 6$ (Fig. A5.8), a loss of message value of $1 \cdot 5\%$ (Fig. A5.6) and an average message delay of 20 seconds before it can be transmitted (Fig. A5.4).

Let us now examine the case of four channels or trunks with four times the number of mobiles as in the single channel case, but with random access facilities.

Reference to Fig. A5.12 shows that, with a busy hour loading of just over 106 mobiles (equal to $4 \times 26 \cdot 6$), the total traffic flow becomes $1 \cdot 6$ Erlangs. Projecting this into Fig. A5.13 and cutting the 4 channel curve we now arrive at a grade of service of just over $0 \cdot 05$ giving one lost call in $18 \cdot 5$ attempts – a vast improvement on the single channel case of 1 in $3 \cdot 6$.

If, however, the grade of service ($0 \cdot 28$) as shown in the single channel is retained, it will be seen from Fig. A5.13 that the total traffic over 4 channels can rise to $3 \cdot 7$ Erlangs which in turn raises the busy hour mobile loading over 4 channels to approximately 250, equal to 62 per channel, an equivalent of 155 mobiles/Erlang.

Thus, with 4 channels or trunks at the specified traffic level, the number of mobiles per channel has been increased by a factor of $2 \cdot 3$ *for the same performance*.

9.   Conclusions

The foregoing notes show the dependency of many factors upon the degree of channel occupancy. Occupancy levels appreciably above about 40% of full loading reduce the grade of service, increase transmission delay and reduce message value, beyond acceptable limits in many cases.

Annexe A.   The Effect of FM Capture on Channel Access

1.   General

The effect of FM capture upon the ability to gain access to a channel is a function of the channel loading and the relative distances from the base station of mobiles requiring entry into the circuit.

Partial obliteration of calls from the weaker mobiles can, in turn, increase the channel loading, further increasing the delay or waiting time.

2.   FM Capture v Grade of Service

Here we look at the grade of service and its reciprocal, the lost call rate.

If it is taken, for instance, that in five random attempts to access a channel, two fail owing to the channel being occupied, it seems equally likely that there is at least a similar chance (two in five) that more than one mobile will be waiting to gain entry to the circuit. The mobiles are therefore likely to call the base station at the same time.

In such a condition, if the mobiles are at markedly different distances from the base station, then with FM signals with capture effect, the distant mobile will be hard-pressed to obtain access.

As channel occupancy is increased, the possibility of lost calls also increases at a rate governed by $(1 + T)/T$ where T is the traffic in a single channel in Erlangs during the busy hour. (Alternatively T can be expressed as a fraction of full occupancy during the busy hour.) Additionally the likelihood of the distant mobile obtaining access to the channel could worsen as the channel occupancy is increased by the extra traffic incurred to offset partial obliteration of calls. Waiting time for entry will therefore be extended.

Furthermore, the effective range of the system could appear to be reduced during the busy periods.

3.   Effect of system configuration

FM capture depends on the equipment design, bandwidth used and signal ratios. If it is assumed that the first two are fixed parameters of the system, then we must examine the probability of the signal ratio between any two mobiles exceeding that required for capture.

The type of system is the main contributor to the probability of $S_2/S_1 > CR$ where S_1 and S_2 are the two mobile signal levels and CR is the ratio of signals needed for adequate capture. The probability is less likely with wide area rural coverage than in an urban system with some suburban or even rural coverage. In the first case, the distribution of mobiles could limit the degree of total capture (mobiles tending to give a more even signal distribution throughout). In the second case the channel loading caused by the large proportion of mobiles likely to be in the urban area could be detrimental to access by those in the rural districts.

Thus, although some modification to the effective grade of service is likely by virtue of mobile distribution, the modification is likely to depend mainly on the system configuration and user pattern.

4. Conclusions

(*a*)   Consider two or more mobiles calling simultaneously, in an environment where differential signal strengths exceed the FM capture ratio needed to suppress the weaker signal. As channel occupancy increases, the probability of mobiles in the weaker signal areas gaining access is reduced and greater waiting times are likely.

(*b*)   The effect of partial obliteration by capture will increase the nominal effective channel occupancy, reduce the grade of signal (increase lost call ratio) and reduce the effective cover of the system during busy periods.

(*c*)   Some improvement in the above effects will take place in systems where the mobile distribution is completely random whilst the worst condition of a heavily loaded centre city core and wide area rural coverage could worsen weak signal capture. Terrain could also modify the general picture. Thus the system configuration enters into the overall picture to a marked extent.

(*d*)   The reduced capture effect in FM equipments with reduced bandwidth produces less obliteration of distant mobiles as a higher S_2/S_1 ratio will be required to achieve complete capture.

Graphs, showing the general effect are included in Ref. 12, pp. 112, 114.

Annexe B. Maximising Loading Efficiency of a Single Channel Throughout the 24 Hour Period

1. General

As indicated by many other authors, there is a maximum busy hour loading level above which the ease of access decreases rapidly. Inevitably, frustration and loss of message value are among the reasons why a radio channel loses its usefulness and why the natural growth of mobile radio can be inhibited.

Where a channel is occupied by a single user, adjustment in operational procedure or the re-arrangement of schedules can often improve the spreading of channel usage throughout the 24 hours. With time sharing of a channel by a number of small users, an optimum mixture of these users can be made so as to even out the long term variations.

The optimum but somewhat Utopian state, where the loading is uniform throughout 24 hours, is obviously virtually impossible but an approach towards this goal is considered desirable and the following paragraphs indicate one method of expressing the degree of optimisation achieved.

2. Busy hour to 24 hour usage ratio. Single User

As we have already seen (Appendix 5), the percentage occupancy is $t \, v \, c/36$, where t = average message length in seconds, v = average number of vehicles using the channel during the busy hour, c = average call rate during busy hour per vehicle.

Observations have shown that, although the average number of vehicles using the channel during the busy hour may vary considerably, an average of 30% of the *licensed* number may then be in use and can be taken as a reasonable figure on which to base the subsequent calculations. The average hourly number of vehicles using the channel over the whole 24 hours is always less than this figure and we will assume 20%.

The average call rate during the busy hour will be a fairly constant percentage of the total 24 hours message content per vehicle, depending, of course, upon the type of business traffic. Averaged over 15 to 20 types of single user channels this figure is around 2 calls per vehicle in the busy hour out of a total of, say, 15 messages per vehicle over the whole 24 hours.

The type of business traffic also influences the average message length. This may vary from as low as 10 seconds to over 60 seconds between different users: the average appears to be about 30 seconds.

Let us consider the formula

$$\frac{T.m}{24 \times 3600} \quad \text{or} \quad \frac{T.m}{864} \,\% \quad \text{or} \quad \frac{n_1.t.m}{864} \,\% \tag{1}$$

from which *average* percentage loading for the total system during any one hour over the 24 hour period can be calculated.

T   =   total message period in seconds occupied by one vehicle during 24 hours. (Using the above example, this is 15×30 seconds or 450 seconds total time.) T also equals $n_1 \times t$.

m   =   average number of licensed mobiles using the channel during the 24 hour period. (From the foregoing remarks this could equal 20% of the total number of *licensed* vehicles which we will assume for this exercise is to be 80 vehicles.)

n_1 =   average number of calls per vehicle during 24 hours.

t   =   average call length.

Let us now express a *figure of merit ratio*, given by

$$\frac{\text{average loading over 24 hrs (\%)}}{\text{average busy hour loading (\%)}} \tag{2}$$

$$= \frac{T.m/864}{t.v.c/36}$$

$$= \frac{T.m}{24 \, t.v.c} \tag{3}$$

This formula can be expressed in slightly different terms:

$$\frac{n_1.m}{24 \, v.c} \tag{4}$$

or

$$\frac{\text{total calls during 24 hrs}}{24 \times \text{total busy hour calls}}$$

Where   c   =   average number of calls per vehicle during busy hour.

v   =   average number of licensed vehicles using the channel during the busy hour.

n_1 =   as before.

These parameters assume a single averaged message length throughout.

Either formula gives a figure of merit of 0.208.

Continuing, we can calculate the busy hour occupancy as

$$30 \times (0 \cdot 3 \times 80) \times 2/36 \; = \; 40\%$$

This has been postulated as acceptable, providing a reasonable holding to waiting time, with grade of service and loss of message value within normal operational tolerances.

The figure of merit shows that the average loading over the 24 hours is well under a quarter of the busy hour figure. A single user can therefore make much greater use of the channel by increased use of the "non-busy hour" period. By such a redistribution, the busy hour figure can possibly be reduced by transferring some of the traffic to other periods. This, of course, allows for extra mobiles, provided the busy hour loading does not thereby exceed 40%.

The result is a smoother occupancy curve with greater channel efficiency.

Even greater improvement will result from users with vastly different patterns sharing the channel. For instance, a (nocturnal) office cleaning organisation can share a channel used normally by day and thus effectively double the channel utilisation.

3.   Busy hour to 24 hour usage ratio – multiple user

This section deals with the situation in which a number of minor users are allocated the same channel in the same area.

The assessment must be based on average figures if the calculation is to remain within reasonable limits and Fig. A5B.1 shows a hypothetical case of six users allocated a common channel.

The chart shows how the busy periods vary somewhat between users. It also indicates differences in the average message length together with other parameters such as call rate and number of vehicles.

By far the simplest method of calculating channel occupancy is to average the factors affected by time (message length and call rate) and to add the quantities of vehicles (both busy hour and 24 hour averages). These various figures, are shown at the right hand side of Fig. A5B.1.

The figures of merit, explained earlier, can be easily calculated by the following formula:

$$\frac{\text{Total number of calls during 24 hour period}}{24 \times \text{Total number of calls during the busy hour}}$$

Thus from Fig. A5B.1 we have $459/24 \times 65 = 0 \cdot 294$

The busy hour occupancy is best calculated by
Average mean message length "t"
    times
Total average busy hour vehicle complement "v"
    times
Average mean call rate "c" (Busy hour)/36

which equals $21 \times 42 \times 1 \cdot 63/36 = 39 \cdot 9\%$

(see Fig. A5B.1)

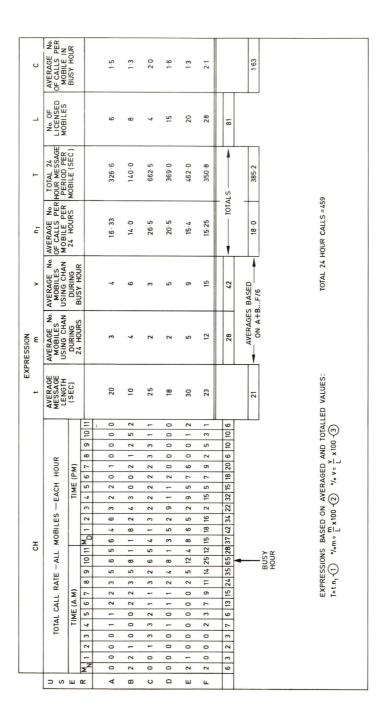

Fig. A5B.1 Summary for 24 hours

7. Conclusions

From the two examples (assuming approximately equal busy hour occupancy) we see a marked increase in the figure of merit with the multiple user case. This demonstrates that, if suitably arranged, the usage efficiency of a channel can be higher in the multiple user case. Indeed, whilst the example was not selected with this in mind, it is evident that careful arrangement of the users could produce quite high figures of merit.

Annexe C. Simplified Channel Occupancy Calculations

1. General

Provided the statistical detail is available, channel occupancy calculations are quite simple and straightforward. They can, however, be tedious to undertake if figures are required at a number of levels; although individual graphs are available, it can be confusing to have to refer to several at one time.

To help evaluate relevant quantities, two composite graphs have been constructed; their method of use is detailed below.

2. Channel occupancy as a function of average call length, average number of calls and mobiles using the channel during the busy hour

Fig. A5C.1 shows a composite graph in which average call length and average number of calls per mobile during the busy hour have been projected to show traffic flow in Erlangs per mobile.

Further projection through the number of busy hour mobiles indicates the channel occupancy as a percentage of total occupancy or directly into Erlangs of traffic flow in the channel. From this figure, the ratio of average waiting to average holding time can be ascertained, this measure based on random attempts to access the channel.

The example shown originates with a call length of 20 seconds. Projecting this value horizontally until the busy hour call rate is intercepted and then projecting downwards at right angles gives a traffic flow of 0·0083 Erlang per mobile for a 1·5 call rate during the busy hour.

Continuing vertically downwards until the busy hour mobile line is intercepted (40 mobiles), an occupancy figure of 33·3% or 0·333 Erlang is indicated. To the right, the same horizontal line meets the waiting/holding time curve at a ratio of 0·5, indicating that waiting times of the order of 0·5 × 20 = 10 seconds can be expected with random attempts to access.

Entry to any point in the composite graph can obviously be made if the necessary values are available.

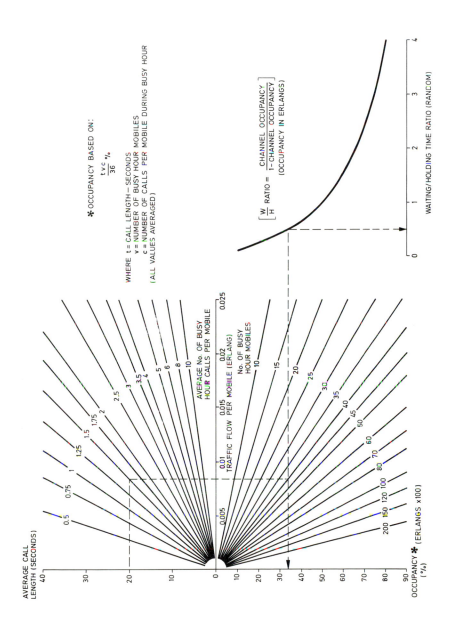

Fig. A5C.1 Relation between Call Length, Occupancy, Traffic, etc.

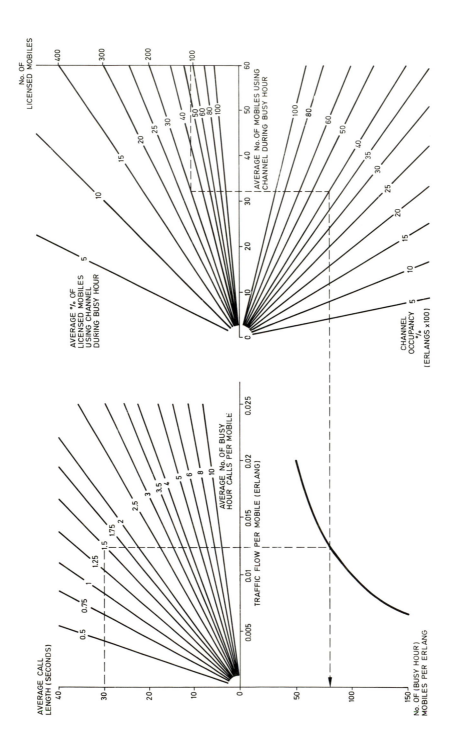

Fig. A5C.2 Relationships Including Percentage of Licensed Mobiles in Use

3. Channel occupancy as a function of average call length, average number of calls and percentage of licensed mobiles operating

Fig. A5C.2 is a development of Fig. A5C.1 involving not only the number of calls and call lengths, but also the percentage of licensed mobiles in use during the busy hour.

In the example, a call length of 30 seconds together with a call rate of 1·5 shows a traffic flow of 0·0125 Erlang per mobile. Continuing downwards until the lower curve is intercepted, we see that, with this traffic flow, there could be a loading of 80 mobiles per Erlang of traffic.

If we now consider 107 licensed mobiles with 30% in use during the busy hour, the right hand top graph shows that this will be equal to 32 mobiles. Projecting the line downwards until it intercepts the horizontal line equal to 80 mobiles/Erlang, we see that the channel occupancy will be 40% or 0·4 Erlang of traffic flow.

As with Fig. A5C.1, this figure can be entered at any point given the required quantities.

Appendix 6. Minimising Intermodulation in VHF/UHF Radiotelephone Systems

1. Introduction

Over the past years the rapid expansion in VHF/UHF radiotelephone use has given rise to more communal siting. In a restricted area, a large number of schemes may be operating on different frequencies. Examples are found, for instance, at many airfield installations throughout the world.

The increased use of VHF/UHF communications in built-up areas has, in turn, introduced many problems of interference. Double frequency operation has minimised unwanted effects, but in certain countries, adherence to single frequency working has created a problem, especially where blocks of adjacent frequencies are used at communal or closely spaced sites.

Although equipment design has reached an advanced state it is inevitable that with further expansion, precautions must be taken to reduce the possibility of interference between schemes operating under adverse conditions. This Appendix shows how intermodulation interference effects may be prevented or minimised.

2. Intermodulation Effects and their Causes

If two signals are applied to a non-linear device, mixing will occur and additional spurious signals will result. The amount of non-linearity and the strength of the wanted signals will, of course, affect the amplitude of the unwanted signal or carrier.

If a block of, say six adjacent frequencies is allocated to one user on closely spaced or communal sites and the condition for intermodulation (common coupling in a non-linear device) exists, then spurious signals will arise. If the allocated frequencies are evenly spaced, and single frequency working is employed, the spurious signals can coincide with other channels in the same scheme. For example, intermodulation between two signals on, say, 80 MHz and 80·1 MHz will produce spurious signals on 79·9 MHz and 80·2 MHz which could well be other channel frequencies in use by the same organisation. These spurious signals are produced by the non-linear device causing harmonics of 80 MHz and 80·1 MHz. The second harmonic of 80 MHz (160 MHz) beats with 80·1 MHz to produce 79·9 MHz and the second harmonic of 80·1 MHz (160·2 MHz) beats with 80 MHz to produce 80·2 MHz.

Additionally, further products can occur by the third harmonic of 80 MHz (240 MHz) beating with the second harmonic of 80·1 MHz (160·2 MHz) to produce 79·8 MHz etc. Fortunately, however, the higher order of harmonics are usually attenuated rather rapidly.

Modulation of both the primary signals will appear on each of the spurious signals. With amplitude modulation, one signal will appear higher pitched than the other, while with a frequency modulated signal, one modulation will be much louder than the other. This is because one of the primary signal frequencies is multiplied in the non-linear device.

2.1 Cause of Intermodulation. These are three possible origins of intermodulation in a system. These are:

(*a*)   In the transmitter output stages and associated equipment (e.g. antennae), caused by the non-linearity of the output stages.

(*b*)   In the receiver input circuits by overloading causing non-linearity of the RF and mixer stages.

(*c*)   In the vicinity of the equipment (usually the transmitter) where metalwork (in masts, buildings etc.) produces what is popularly known as the "rusty bolt" effect. This is caused by bad metallic contact between rusty or corroded metals (or possibly by dissimilar metals in contact) producing a rectifying action when in a strong RF field. The resultant non-linearity will set up spurious signals in the surrounding metalwork which can radiate the unwanted carrier.

2.1.1 *Transmitter Intermodulation.* The main cause of intermodulation between transmitters on closely-spaced channels is their close coupling. Coupling mainly exists in the antenna system, and as a common mast is often used, the coupling between antennae can be very tight unless adequate care is taken. The effect of this coupling is to feed voltages from one transmitter to another. As output stages operate invariably in Class C, the non-linearity can be considerable. To reduce intermodulation due to this cause in installations with vertical polarisation, some staggering in antenna height is often much better than horizontal spacing. For example, the attenuation between antennae on a common horizontal plane at a distance of 3 m (10 ft) may be 20 dB while a vertical separation (E-plane) could give almost 40 dB for the same distance. For horizontal polarisation the effect is similar when antennae are mounted end on to one another (E-plane). Transfer of some of the antennae to another mast is recommended where their number is great.

Some intermodulation can occur owing to the closeness of equipment in racks. The possibility of standing waves on feeders in close proximity to one another must not be overlooked, but the majority of intermodulation effects at the transmitters occur by coupling in the antenna system. For this reason, common antenna working on close, evenly spaced, frequencies is possible but not easy, although on channels separated by about 2 or 3% of the nominal frequency it is practised with more success. However, even assuming that common antenna working is practical on two widely separated frequencies, not only is some power often lost in the process of combining, but some intermodulation can still occur on channels not under the control of the offending system unless a complex array of filters and circulators is used. Thus, while satisfactory to the user, common antenna working may cause interference to other users in the vicinity.

2.1.2 *Receiver Intermodulation.* This type of intermodulation is extremely common. In countries employing single frequency simplex, two transmitters in the vicinity of the receiver having a frequency relationship as described in Section 2 can cause intermodulation by overloading the RF section. The effects on both the RF stages as well as the mixer can be non-linearity of a high order, even at distances up to several kilometres. Hence two-frequency working is recommended wherever possible. This ensures that fixed station transmitters in the immediate vicinity are on frequencies separated from the receiver frequencies by an adequate guard band of several MHz.

In initially setting up a single frequency scheme with a number of adjacent channels (e.g. airfield control systems), the antenna spacing (transmitter to receiver) must be very carefully planned and a spacing of not less than 1·6 km (1 mile) is recommended. With this degree of spacing, a 50W transmitter should not seriously affect a receiver on an adjacent channel.

2.1.3 *External Effects.* The main cause of intermodulation in this category is the use of a metal mast for multiple antenna installations on closely spaced frequencies. It is extremely difficult to ensure a perfect metal-to-metal joint at the many points of contact present in masts, especially of the lattice type. The best that can be done with multiple installations of this kind is to clean and thoroughly secure the metal-to-metal joints. Subsequent corrosion can then be reduced by coating with a suitable paint. Alternatively, bonding points of contact by means of copper strip may minimise the trouble. Further causes of intermodulation can be traced to similar effects in nearby guys and metal structures.

There are two ways in which the effects can be avoided. The first is to separate the antennae in such a manner that spurious signals are not only reduced but cannot fall on the other channels in the system. For example, in a four-channel scheme on, say, 80, 80·1, 80·2 and 80·3 MHz, separating the 80 and 80·2 MHz antenna from the 80·1 and 80·3 MHz antenna, will not only reduce the intermodulation (due to the resultant lower local RF field) but any spurious signals will fall on other channels (79·8 MHz, 79·9 MHz, 80·4 MHz, 80·6 MHz) which may be unused, at least in the vicinity of the station.

The second method is recommended for reducing the remaining inter-modulation interference, regardless of the cause. It is to stagger the allocated channels by uneven amounts. The avoidance of frequencies affected by third order interference products which occur when $a + b - c = d$ or $2a - b = c$, will eliminate most of the problems. In the extreme case where fifth order interference is troublesome, further staggering of the frequencies will minimise the effect but only at the expense of frequency band utilisation.

Taking the normal band of $\pm 0\cdot 2\%$ of the nominal frequency over which crystal switching may be achieved with current equipment, it can be seen that, at 150 MHz, a total of 25 channels each 25 kHz wide, is available. If these channels are numbered from 1 to 25, then to avoid third order intermodulation interference falling on the operating channels, only channels 1,2,5,11,19 and 24 can be utilised.

Similarly at 75 MHz where only 13 channels each 25 kHz wide are possible, only channels 1, 2, 5, 10 and 12 can be used.

The following table shows in greater detail, some combinations of operating channels having no third order intermodulation interference.

Channels required	Channels available	Operating channels having no 3rd order intermodulation interference
3	4	1,2,4
4	7	1,2,5,7
5	12	1,2,5,10,12
6	18	1,2,5,11,13,18
7	26	1,2,5,11,19,24,26
8	35	1,2,5,10,16,23,33,35
9	46	1,2,5,14,25,31,39,41,46
10	62	1,2,8,12,27,40,48,57,60,62

In extreme cases it may be necessary to eliminate third and fifth order intermodulation interference from the operating channels and a total of 137 channels for 8 operating channels is required to achieve this. The operating channels will be 1,2,8,12,27,50,78,137. From this it can be seen that if twelve channels are available (at 75 MHz), elimination of both third and fifth order intermodulation products can be achieved by selecting only the first four channels, 1,2,8 and 12. It must, however, be emphasised that intermodulation should be minimised at the source wherever possible before considering the staggering of channels. The staggering of channels is intended to ensure that any remaining intermodulation does not fall on operating channels.

Other methods of staggering can be adopted and the reader is referred to a paper by Edwards, Durkin and Green (Ref. 8).

3. Use of Filters

In the following cases, the use of filters is justified.

(*a*) If the separation between the frequencies causing the intermodulation and the resultant spurious signal is of the order of 2 to 3% and it has been proved that the effect is being produced in the receiver, then a filter on one of the offending frequencies (usually the frequency further from the wanted channel) is usually effective.

For example, if two nearby transmitters on 120 MHz and 121·5 MHz are causing a spurious signal to appear on a receiver tuned to 118·5 MHz by overloading, then a filter rejecting 121·5 MHz at the receiver input may cure the intermodulation. Alternatively, a sharply tuned filter accepting 118·5 MHz only may be preferable in the receiver input. The type of filter to use depends entirely upon the system, as it is obviously out of the question to use a sharply tuned acceptor circuit when a number of channels are to be received on one equipment (e.g. on a switched channel receiver).

(*b*) If intermodulation is occurring at a transmitter site where simultaneous transmission of several frequencies is likely (e.g. a commercial VHF site), then some reduction in intermodulation is possible by using sharply tuned acceptor filters on the offending transmitters. For example, a VHF site may be used by several authorities and the frequencies could be:

Transmit (i) 77 MHz (ii) 79 MHz (iii) 80 MHz (iv) 81 MHz
Receive   (v) 71 MHz (vi) 72 MHz (vii) 73 MHz (viii) 74 MHz

Combinations of (i) and (iii) can produce an intermodulation product on (viii) whilst combinations of (i) and (iv) may affect (vii). If the effect is obviously in the transmitter (this can be proved by trying the effect of an acceptor filter in the receiver), then acceptor filters in (i), (iii) and (iv), or possibly only (i) may eliminate the trouble.

If closely spaced channels are troublesome, the use of isolators will be of assistance (see Appendix 10).

(c) If filters in either the receiver or the transmitter are ineffective it may be that intermodulation is occurring in nearby metalwork. Action can then be taken accordingly.

Summarising, intermodulation effects can be minimised by the following:

(a) In the receiver: by ensuring that signals likely to cause intermodulation are of a reasonably low level. This is usually achieved by double frequency working with adequate spacing between the transmit and receive frequencies. A suitable filter can be of value. With single frequency working on sites accommodating a number of adjacent channels, a transmitter to receiver antenna spacing of at least 1·6 km (1 mile) is recommended.

(b) In the transmitter by suitably spacing antennae either on separate masts or by adjustment of the antennae on the mast. Separation of feeders can also assist in the reduction of intermodulation. As in the receiver, a filter can be useful. On closely-spaced channels isolators can be of value (see Appendix 10).

(c) Externally: by bonding mast joints, separating the antennae on different masts and avoiding the close proximity of metalwork likely to be troublesome. Filters will not assist when this effect is present.

(d) Generally: by using frequencies with uneven separation so that intermodulation products fall off-channel.

4. The Siting of VHF/UHF Transmitters and Receivers and the Allocation of Suitable Frequency Spacing to Avoid Interference

4.1 General. Many interference problems are brought about by overloading of the early stages of receivers by nearby transmitters. The problems of siting and frequency allocations are of major importance in the engineering of systems. The following notes outline some of the precautions necessary, and should be read in conjunction with the previous sections.

4.2 The Problem. In present day receivers, the majority of the final selectivity is achieved in the intermediate frequency amplifier section by the use of "block" or crystal filters.

In the stages preceding the IF amplifier, the natural selectivity is always much wider because, firstly, the tuned circuits operate at VHF or UHF, and secondly, there is a requirement for multi-channel switching over a limited band. A typical bandwidth of the front end of a standard VHF receiver operating at 100 MHz may be about 20 dB down at ± 1 MHz from the operating frequency.

A transmitter in the immediate vicinity radiating on a frequency within the bandwidth of the receiver early stages, could overload the RF or mixer stages of the receiver to the extent of producing gross non-linearity. Blocking, cross-modulation and other effects would result and the complete system would be degraded or could even become unworkable.

4.2 The Basic Cure. The level of input signal at which non-linearity becomes apparent can be determined. With the level known, it is a simple matter to determine the parameters to avoid overloading the receiver. It is assumed that the effective transmitter radiated power is known.

To achieve the necessary protection to avoid receiver overloading by a particular transmitter, both frequency spacing and physical spacing must be examined. The maximum signal at the receiver input caused by any offending transmitter should be approximately equivalent to an on-frequency signal of less than approximately –50 dB relative to 1 W if blocking and cross-modulation is to be avoided, and less than about –70 dB relative to 1 W to minimise intermodulation effects.

These protection figures can be obtained by frequency spacing between the transmitter and receiver, physical spacing between the transmitter and receiver antennae (and equipment) or a combination of both. It is, therefore, necessary to separate the transmitter and receiver sites when operating in the single frequency mode on adjacent channels, and to specify the minimum frequency spacing requirement for a duplex channel with co-sited equipments.

4.3 Protection Achieved by Physical Spacing between Antennae. In most systems, the antenna layout is such that the advantages of exact end-on or E-plane mounting cannot be realised. Thus the distance/attenuation curve lies approximately between free space and twice free space. Typical attenuation figures between antennae erected in this manner are:

> 6·1 m (20 ft) spacing: 29 dB
> 62 m (200 ft) spacing: 49 dB
> 610 m (2000 ft) spacing: 79 dB

If, however, true E-plane mounting can be achieved so that one antenna lies in the null of the other, then 50 dB would be possible at 6.1 m (20 ft) with 72 dB available at 62 m (200 ft). Normally it is necessary to adjust for minimum coupling between antennae if this method is adopted, but it must be realised that the method can only be used for a limited number of antennae on any one mast and is, of course, a function of frequency.

4.4 Protection Achieved by Frequency Separation between Transmitter and Receiver. The normal curve giving protection versus frequency spacing follows closely that showing the front-end selectivity of the receiver. Obviously there are variations in selectivity between equipment of different types and the approximate protection would be between 3 and 8 dB at 0·5% off tune, 10 and 25 dB at 1% off tune, 25 and 60 dB at 2% off tune and, 50 and 95 dB at 5% off tune.

At spacings of less than 0·5 MHz at VHF or UHF the protection is negligible, and therefore equipments operating on adjacent channels must obtain all the protection by physical spacing only.

4.5   Coupling between Transmitter and Receiver other than by Antennae. When the frequency spacing is marginal, and protection by antenna spacing is envisaged using minimised end-on coupling of antennae on a common mast, care must be taken to ensure that there is no breakthrough or coupling between equipments in the same rack or even in the same building, or between feeder runs. Any such coupling will reduce the required protection.

4.6   Practical Examples.

(*a*)   Two 50W transmitters on the marine band are operating on Channels 12 and 14. It is required to listen through on Channel 16. What is the physical spacing required between transmitter and receiver sites, taking into account:

(i)   Blocking and cross-modulation effects only (Channel 12 or 14 operating, one at a time), and

(ii)   The possibility of intermodulation at the receiver (due to both Channels 12 and 14 transmitting simultaneously)?

(i)   If the blocking and cross-modulation level $= -50$ dB relative to 1W, the protection required will be $50 + (10 \log_{10} 50)$ dB $= 50 + 17$ db $= 67$ dB. No protection is available in receiver selectivity, therefore physical spacing must be greater than 300 m (1000 ft) to achieve 67 dB.

(ii)   If the intermodulation level needed is -70 dB relative to 1W, the protection required will be $70 + (10 \log_{10} 50)$ dB $= 70 + 17$ dB $= 87$ dB. No protection is available in receiver selectivity, therefore physical spacing must be greater than 1036 m (3400 ft) to achieve 87 dB.

(*b*)   What is the necessary frequency spacing to operate a talkthrough (or duplex) system using a 50W transmitter co-sited with the receiver both in the 160 MHz band with antennae 6·1 m (20 ft) apart, assuming no coupling between equipment?

The required protection will be $50 + (10 \log_{10} 50)$ dB $= 50 + 17$ dB $= 67$ dB. The protection achieved by the antenna spacing will be 29 dB. Therefore the frequency spacing must be adequate to give $67 - 29$ dB $= 38$ dB protection. The minimum frequency spacing to achieve 38 dB isolation would lie between 4·2 and 6 MHz according to the type of receiver used and antenna filters fitted.

4.7   Summary. Protection against blocking, cross-modulation, and intermodulation of the receiver can be achieved in normal VHF/UHF systems only by ensuring that the receiver is not subjected to excessive voltages at unwanted frequencies. This protection can be achieved by physical spacing, frequency separation or by a combination of both.

5.   Siting of Transmitting and Receiving Stations in Single and Two Frequency Systems with Particular Reference to International Maritime Services

5.1   General. The International Maritime Services, owing to the use of single and double frequency channels, is one of the best examples of the importance of the correct siting of transmitters and receivers to avoid interference and spurious effects. The satisfactory installation of a maritime scheme can only be achieved by care in eliminating or minimising interference. The following notes refer to the maritime service in particular, but the reader should also refer to the points raised earlier when considering other types of system.

5.2   Shore Station Siting. General Requirements. In many instances, the shore stations in maritime systems are located within short distances of ports or shipping areas where the concentration of traffic is high. Furthermore, there is often a requirement for listening through to all channels allocated to a particular area. There is also a risk of the overloading of receivers by simultaneous transmission on adjacent channels from the shore transmitters. The close proximity, both to the shore station and to other vessels, of ship-borne transmitters raises the problems of intermodulation, blocking, etc. already described.

To overcome or minimise these effects, the following must be considered:

5.2.1   *Shore Receiving Equipment.* The receivers must be separated from their companion transmitters by a minimum distance, determined by frequency spacing and transmitter power, to permit listening through to all other channels allocated to the area, even during transmission on adjacent channels. In addition, the minimum spacing must be determined by the possibility of intermodulation or, alternatively, whether intermodulation might arise if further channels are added to the system.

Paragraphs 1 to 3 cover this aspect fully, and it sufficient to state that, since marine frequencies are equally spaced, they are therefore prone to on-channel intermodulation. For example, Channels 12, 14 or 16 can produce on-channel intermodulation interference when Ch 12/14 or Ch 14/16 radiate. Likewise, Channels 6, 8, 14 and 16 will give likely on-channel intermodulation interference (based on $a + b - c = d$) if three transmitters operate simultaneously.

On the other hand, even if on-channel intermodulation effects are not present, the blocking of wanted signals when an adjacent channel transmitter radiates, can upset the scheme by desensitising the receiver during these periods.

The cure for the problem, therefore, can be simply siting the shore receiver station sufficiently far from its companion transmitter to avoid overloading, while at the same time ensuring that the site chosen is sufficiently remote from the nearest vessels to avoid the transmitters on those vessels producing the same ill-effects.

When using 25W shore transmitters, a spacing of 1·6 km (1 mile) between the receivers and transmitters, and also between the receivers and the nearest vessels, is recommended to minimise intermodulation effects.

If the allocated channels cannot give on-channel intermodulation interference, this minimum distance can be reduced to 400 m (¼ mile). This distance, covering only blocking, must be decided upon only if there is no possibility of further channels being added to give rise to intermodulation combinations later.

5.2.2   *Shore Station Transmitting Equipment.* The only other point to be considered when siting transmitting stations is the possibility of producing intermodulation and blocking effects in nearby ships' receivers.

It is obviously ideal to locate the transmitters so that the distance between them and ships is of the same order as between the transmitters and their companion receivers, but usually this is impracticable owing to control restrictions. "Low power" switching tends to reduce the effects and it is usually appreciated that vessels within, say, 300 m (1000 ft) of the transmitters will be affected to some degree. They are unlikely, however, to be communicating over distances where blocking of the receiver will be of importance, so that any interference is not likely to be excessive until the ships' receivers are very close to the shore transmitters.

5.3 Effect on Ships' Equipment. There will be a number of occasions when vessels operating on the International Maritime Bands in port will be subject to interference from, and will be the cause of interference to other vessels. It is assumed that the shore receiving station is sufficiently far from the vessels not to be affected by them.

If, therefore, a number of vessels within a dock area are operating on several port operation channel frequencies, say 6, 8 and 10, then under conditions of receiver overload, both blocking (desensitisation of receivers) and intermodulation could occur. Usually the distance between any two vessels operating on a port operational channel is relatively short, so that while interference will be present (and, of course, very severe during listening-out periods) the level during reception of a strong on-channel interfering signal will be modified by the much stronger wanted signal present, and therefore the full interference effect may not be so marked. Because of the various possible levels of signal, the distance at which interference will become intolerable cannot be defined and only experience in a particular port area will determine the severity of the interference.

Some degree of control by the port shore station and judicious use of the "low power" facility will reduce the ill-effects to a minimumm.

5.4 Summary. The following points summarise the preceding details:

  (a) The shore receiving station must be located at least 400 m (¼ mile) from both the transmitting station and the nearest vessel concentrations, to avoid blocking (desensitisation of the receivers) where the shore transmitters radiate not more than 50W.

  (b) The distance must be increased to at least 1 km (0·6 mile) and preferably 1·6 km (1 mile) if, because of the frequency allocation, "on-channel" intermodulation interference is possible. It is fairly safe to say that, in most ports, this situation occurs.

  (c) The transmitting site should also, if possible, be 400 m (¼ mile) to 1·6 km (1 mile) from the nearest vessel concentration, to minimise interference to ships' receivers. This distance cannot, in many cases be achieved owing to control reasons, but where only one shore operator is envisaged, only one channel may be radiating at a time, thus avoiding the possibility of intermodulation.

  (d) Some interference between ships is likely in areas where several groups are operating simultaneously. The degree of interference cannot be calculated because many factors are involved, but the effects should be tolerable in all but extreme conditions. Control by the port shore station usually reduces the effect to a minimum.

(*e*)   Listening through on the single frequency simplex channels is normally not possible on any one vessel because of the lack of separation between transmitters and receivers.

Annexe A. Intermodulation – An Explanation of the Various Products

1. General

Intermodulation is caused by a number of effects, the mechanism of which is described in detail in Appendices 7 and 9. The effect on a receiver depends on the "order" of interference which can be produced at the source.

2. Category of Interference

Intermodulation products fall into a number of categories or orders as follows:

2.1 Second Order Products. These are based on the sum or the difference of two frequencies (a ± b). Thus, for example, if two frequencies 75 MHz and 80 MHz are present in the vicinity of the affected receiver, then second order products at 5 MHz and 155 MHz may exist.

Second order interference does not usually produce high levels of interference, for the simple reason that the unwanted products are generally many megahertz removed from the working frequency. The selectivity of the receiver "front end", and of the transmitter output stages, usually provide attenuation of the unwanted products and, if any appreciable level is left, the additional attenuation afforded by an extra filter is usually more than sufficient to remove the remainder.

However, interference can occur in the vicinity of high power broadcasting stations in the MF, HF and VHF bands where the difference product between a MF or HF station and a local VHF station falls on the wanted VHF channel. For example, a VHF broadcasting station at 92 MHz and a high power HF transmitter at 10 MHz could produce a high level of interference in the region of 82 MHz. This interference could extend over several communication channels because of the 75 kHz deviation in the case of the VHF broadcasting station.

2.2 Third Order Products (2a – b; a + b – c)*. This order is probably the easiest to produce and the most difficult to eradicate.

The main reason why problems occur with this product is the ease with which the interference gains access to the vulnerable parts of the affected circuit. This is due to the fact that all the frequencies (the two or three causing the problem and the affected frequency) can often lie close to one another. Thus tuned circuits often exert little or no effect.

A typical example is given by the following three frequencies:

  (i)   80 MHz
  (ii)   80·1 MHz
  (iii) 80·2 MHz

With these frequencies, third order products could fall at 79·8 MHz, 79·9 MHz, 80·0 MHz, 80·2 MHz, 80·3 MHz and 80·4 MHz. These cannot be cleared by normal filters and two contradictory features, namely high selectivity and channel switching, do nothing to help solve the problem in the first instance.

2.3 Fourth Order Products (3a ± b; 2a ± 2b; a ± b ± c ± d etc). These products have the same character as those for second order – namely, products removed by a considerable amount from the source frequencies. Under these circumstances, the elimination of interference is relatively easy compared with that of odd orders.

2.4 Fifth Order Products (3a – 2b; 2a – 3b etc.)*. These products follow the general trend of third order products – namely close frequency interference – and are, similarly, difficult to eradicate.

However, being of higher order, the amplitude is usually lower than with third order effects and, thus, the interference is less.

2.5 Higher Even Order Products. These follow the previous even order products but, being of even lower amplitude, the effects are likely to be negligible.

2.6 Higher Odd Order Products*. These will decrease in effect rapidly as the order increases and seventh order and above are unlikely to be troublesome.

---

*(It should be noted that odd order products also occur when the primary frequencies are added, for example, 2a + b, 3a + 2b, 2a + 3b. However, these products fall well out of the band in use and therefore normally have no perceptible effect.)

Annexe B.   The Elimination of a + b – c On-Channel Intermodulation Products in Duplex or Talkthrough Systems

1. General

The odd order on-channel intermodulation products which are based on the mixing of two frequencies out of a number can be relatively easily removed "off channel" by uneven spacing of the frequencies allocated to the units located on the site. The process is detailed earlier in the Appendix.

However, where a high level signal from a mobile might cause non-linearity in a fixed receiver and, at the same time, the base transmitter of that channel is radiating, then the receiver of a second channel could experience intermodulation interference when its own transmitter radiates.

The following paragraphs outline the problem in greater detail and indicate a method of avoiding the likely effects.

2. Interference Mechanism

It is usual for frequency allocations to be based on identical channel spacings in the transmit and receive blocks. By using irregular spacing as outlined earlier, on-channel third order intermodulation involving $2a - b$ products can be avoided.

However, with the $a + b - c$ product, identical transmit to receive channel spacing does not move the interference off the channel concerned.

For instance, consider two duplex channels with transmitters on f_A and $f_A + \Delta f$ and receivers on f_B and $f_B + \Delta f$, with Δf being the same in both cases.

Then, if both transmitters are radiating and a strong mobile signal is received on one channel,

$$f_A \quad + \quad (f_B + \Delta f) \quad - \quad (f_A + \Delta f) \quad = \quad \text{frequency subjected to interference}$$

$$\underbrace{}_{a} \qquad \underbrace{}_{b} \qquad \underbrace{}_{c} \qquad \underbrace{\phantom{\text{frequency subjected to interference}}}_{d}$$

fixed	mobile	fixed
transmitter	transmitter	transmitter

cancelling throughout, $d = f_B$.

3. Detrimental Effects caused by On-channel Intermodulation Products a + b − c

3.1 Both Stations Operating in the Talkthrough Mode with the following conditions·

(*a*) Both transmitters triggered, f_A and $f_A + \Delta f$.
(*b*) Mobile transmitter on $f_B + \Delta f$ in the vicinity of the repeater station site.
(*c*) Mobile transmitter on f_B at any distance from the repeater station site.

The spurious intermodulation response will be generated on f_B. This may or may not be at a higher level than the mobile signal being received on f_B.

(i) If the mobile signal level on f_B is higher, then the intermodulation signal will not be heard *until* the f_B mobile signal is removed. Then the talk-through facility on this channel will be locked on until the strong mobile signal transmitting on $f_B + \Delta f$ is switched off.

Both talkthrough systems will then revert to the quiescent state.

(ii) If the mobile signal level is lower than the spurious intermodulation response, the latter will capture in an FM system, or take over by signal level in an AM system. So the system will be locked until the mobile transmitter on $f_B + \Delta f$ is switched off, or the level from it drops below that necessary to produce the intermodulation response.

3.2 Both Stations Operating in the Duplex (telephone) Mode.

(*a*) Where both circuits are operating to simplex outstations (either fixed or mobile), the generation of a spurious intermodulation product on f_B when a strong mobile signal exists on $f_B + \Delta f$ will cause unwanted interference to the *line* subscriber when he is speaking (i.e. when the mobile is receiving).
(*b*) When the signal from the mobile on f_B is at a lower level than the intermodulation product, then severe interference or capture by the product will occur. This will result in the circuit being unusable.
(*c*) Where both circuits are operating in the full duplex mode, then, under the conditions given in (*b*), the lower frequency channel will be rendered unusable by interference at all times.

4. Avoidance of the On-channel Effect by the Use of Irregular Spacing

It can be seen from the equation that the most obvious method of avoiding an on-channel combination is to make Δf different in the case of each transmitter and receiver. This is difficult to achieve where a considerable number of systems are operational in the same band on the same site, but it is relatively easy where the total number of channels is small.

Fig. A6B.1 shows an example based on the '6 from 18 channels' third order on-channel intermodulation-free sequence and, assuming two blocks are available, the $a + b - c$ effect will be eliminated when any two of three duplex channels operate.

With any organisation having a number of adjacent channels available for its own use, the irregular spacing of channels should be no problem and the suggested method of avoiding on-channel interference would be exercised over a number of sites whilst still not exceeding the allocated block.

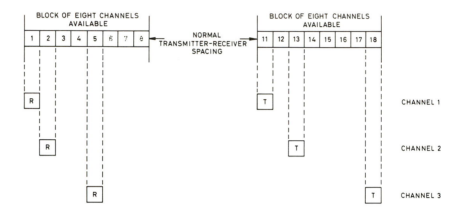

Fig. A6B.1 Third-order On-channel Intermodulation – free Sequence

Annexe C. Identification of Odd Order Intermodulation Products by Matrix Method

1. General

The identification of intermodulation products and in particular those of the higher orders is often a prolonged and monotonous task. Aids to simplify the process have been produced and these range from graphical means to full computer analysis.

Undoubtedly a computer can eliminate a vast amount of cross calculation and when available with an adequate program it should be used in the more complex cases. However, the use of a simple matrix in the less complicated cases can generally produce the desired answers fairly quickly and more cheaply.

The following paragraphs detail the proposed matrix method.

2. Matrix Construction

Fig. A6C.1 (i) shows the matrix drawn to accommodate eight frequencies or channel numbers. Either parameter can be used provided the same terms are used throughout. In the case of channel numbers, the spacings between individual channels must be identical. The number of discrete channels or frequencies shown on a matrix can, of course, be reduced or increased as desired.

(a) The first requirement is to arrange the channels or frequencies in ascending or descending order in the left hand column.

(b) The next column (first difference) is completed to show the differences between adjacent channels or frequencies.

(c) Column three (second difference) shows the differences between the first and last of any group of three consecutive channels or frequencies.

(d) The fourth column (third difference) indicates the difference between the first and last of any group of four consecutive channels or frequencies.

(e) The process is repeated until the matrix is complete.

Reference to Fig. A6C.1 (ii) indicates a simple method of compiling the matrix:

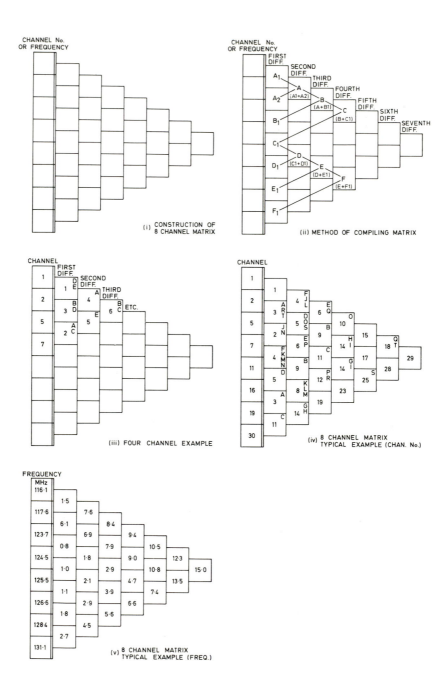

Fig. A6C.1 Matrix Construction

(i) The first difference is the difference between adjacent frequencies or channel numbers.

(ii) The second difference is the sum of the first differences overlapping the required "square".

For example –
A = A1 + A2
D = C1 + D1

(iii) The third and subsequent differences are the individual sums of previous differences and the next first difference (ascertained by projecting the difference columns diagonally).

For example –
B = A + B1
C = B + C1
E = D + E1
F = E + F1
etc.

Fig. A6C.1 (iii), (iv) and (v) show completed matrices – two based on channel numbers and the third on actual frequencies.

3. Analysing the Matrix

Analysing the matrix is simply a process of reading the figures and noting any basic whole number ratio between the figures in the squares. Obviously, channel numbers tend to be simpler to read and such a method of identifying the main signals may be preferred. However, conversion from frequencies to channel numbers could well be a prolonged process, requiring some care to avoid misplacing the sequence, and therefore the use of frequencies may be safer where they have been supplied originally.

3.1 Explanation of Order Identification. This can be derived from a simple equation employing the whole number ratio of the differences. Obviously when two differences are identical a third order product is present; a ratio of two to one in the differences will produce a fifth order product; a three to one difference, a seventh order product and so on.

The whole number ratio can therefore be deduced from the following expression for the order of the intermodulation product.

$$\text{order} = \frac{2y}{x} + 1$$

Where y is the larger of the two differences
    x is the smaller of the two differences
and   y/x is a whole number ratio.

However, where the smaller number of the ratio is contained within the larger number and a common channel exists, then the calculated order must be based

$$\text{on} \quad \frac{2y}{x} - 1 \quad \text{and not} \quad \frac{2y}{x} + 1$$

For example, with Channels 1, 5 and 7

$$= \quad \text{fifth order} \quad \frac{2y}{x} + 1$$

$$= \quad \text{fifth order} \quad \frac{2y}{x} - 1 \quad \text{not seventh order}$$

3.2 Example A (Fig. A6C.1 (iii)). This is a simple case and the channel numbers are those recommended for a four channel case with no on-channel third order intermodulation.

It can be seen that within the difference columns no two differences are the same. This indicates that no third order on-channel intermodulation exists.

There are however a number of combinations showing a whole number ratio and these are tabulated as follows with letters indicating the relevant combinations.

		Order	Whole Number ratio
A	Channels 1,5,7	5th	4 : 2
B	Channels 2,5,1,7	5th	6 : 3
C	Channels 1,2,5	7th	3 : 1
D	Channels 1,2,7	11th	5 : 1

3.3 Example B. Examining Fig. A6C.1 (iv) (an extremely bad case), it can be seen that a considerable number of whole number ratios exist within the matrix and some of these have been identified by letters. There are many more than those shown but the principle of identification is the same throughout the analysis.

		Order	Whole Number ratio
A	Channels 2,5,16,19	3rd	3 : 3
B	Channels 2,11,7,16	3rd	9 : 9
C	Channels 5,16,19,30	3rd	11 : 11
D	Channels 2,7,11,16	3rd	5 : 5
E	Channels 1,7,5,11	3rd	6 : 6
F	Channels 1,5,7,11	3rd	4 : 4
G	Channels 5,19,16,30	3rd	14 : 14
H	Channels 2,16,30	3rd	14 : 14
I	Channels 2,16,5,19	3rd	14 : 14
J	Channels 1,5,7	5th	4 : 2
K	Channels 7,11,19	5th	8 : 4
L	Channels 1,5,11,19	5th	8 : 4

M	Channels 7,11,19	5th	8:4
N	Channels 5,7,11	5th	4:2
O	Channels 1,11,2,7	5th	10:5
P	Channels 5,11,7,19	5th	12:6
Q	Channels 1,7,19	7th	18:6
R	Channels 2,5,7,19	9th	12:3
S	Channels 2,7,5,30	11th	25:5
T	Channels 2,5,1,19	13th	18:3

etc., etc.

3.4   Example C. Fig. A6C.1 (v) can be treated in the same manner as the previous two examples although the differences are in frequency and not channel numbers. The breakdown of the various orders has not been included but can be analysed in the same manner as the previous examples.

4.   Use of the matrix for determining multiple source products

With communal sites where individual channels are heavily loaded, there is a high chance of more than three transmitters radiating simultaneously. Products will be generated based on the addition and subtraction of certain of the "squares" within the matrix and a typical example involving Fig. A6C.1 (iv) is as follows:

(a)   Ch. 30 – Ch. 19 (Difference 11)
      equals
        Ch. 19 – Ch. 16 (Difference 3)
      + Ch. 16 – Ch. 11 (Difference 5)
      + Ch. 7 – Ch. 5 (Difference 2)   Difference total = 11
      + Ch. 2 – Ch. 1 (Difference 1)

If this combination is rationalised, we have, for instance, one combination giving:

Ch. 30 = 2 × (Ch. 19) – Ch. 11 + Ch. 7 – Ch. 5 + Ch. 2 – Ch. 1

Adding the number of "effective" channels together, we have a total of seven, indicating a seventh order product derived from six channels radiating simultaneously. Other combinations from this basic arrangement can also be seen to occur.

Combinations involving other channels can also be found, for instance, within the example typically:

Ch. 1 = Ch. 2 – Ch. 5 + Ch. 7 + Ch. 16 – Ch. 19

(A fifth order product involving the simultaneous radiation of five channels.)

5.   Degree of Importance of Various Orders

Obviously the lower the order of on-channel intermodulation the worse will be the effect.

Most of the highest orders can be ignored and only in extreme cases is it ever essential to analyse the products beyond the fifth order.

With communal sites, third order on-channel intermodulation must be completely eliminated and also possibly the fifth. Only in remote cases is the seventh order likely to be of any trouble.

With single frequency simplex operation as on closely spaced sites, some care is needed to ensure freedom from interference and the matrix method enables a rapid assessment of the higher orders to be obtained where needed.

Annexe D.   Analysis of the Number of Third Order Products in Multi-channel Systems

1.   General

Before considering the ways in which the intermodulation products can be minimised or rearranged to eliminate on-channel interference, it is as well to visualise the quantity of products likely to occur in certain types of system.

With a number of frequencies (either adjacent or within limited area of spectrum), one is faced with two categories of third order products. They are those generated both in and outside the portion of spectrum in use on the site and those generated within that portion of spectrum and likely to affect the operation of systems operating on that site.

Products generated on the site or in nearby receivers can, under unfavourable conditions, affect other adjacent systems but outside the band. It is essential, therefore, to appreciate the magnitude of the problem and reduce the levels of the offending products as well as resorting to the recommended irregular frequency spacings.

The following sections show the quantity of third order products possible in multi-channel systems both outside and within the band in use.

2.   Third order products of 2a–b combination

Within systems having a random distribution of transmitters and a low to medium value of average channel occupancy, the chance of two transmitters radiating simultaneously is quite high. The possibility of three units operating simultaneously is lower but increases quite rapidly as the channel loading rises.

It is usual, therefore, to consider the 2a–b type of product as more likely in these systems. However, the possibility of any other combination cannot be ignored, particularly in systems with higher channel loading.

In the following analysis the worst case is considered – that of adjacent channels – with illustrations of the number of possible 2a–b products that can arise both outside and within the band used.

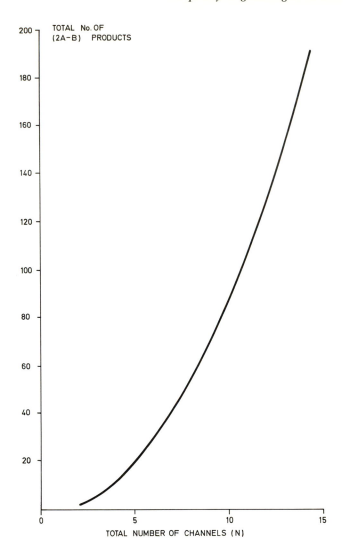

Fig. A6D.1 Generation of (2A–B) Products Relative to Number of Channels

2.1   Total number of products generated. Referring to Fig. A6D.1, the curve illustrates clearly the rate at which the 2a–b products are generated as the number of channels increases. In this context it does not matter whether the channels in use are adjacent or separated, the resultant products are the *total* number of 2a–b products generated. A simple formula (N(N–1)) can be used to derive the number possible using any particular number (N) of channels.

It is interesting to note that higher order products, such as fifth or seventh, can also be obtained from this particular graph. The rate of increase of these products, each involving two transmitters, increases at the same rate as that of the third order products.

2.2   2a–b products generated in-band. On any site with a number of systems, the first essential is to ensure that satisfactory operation is possible by eliminating on-channel intermodulation products.

We have seen earlier in the Appendix how this can be achieved. It is however, important to consider the number of possible 2a–b combinations which can occur in the band if equi-spaced adjacent channels are used instead of irregular spacing.

Fig. A6D.2 illustrates the rate of increase in 2a–b products falling in-band as the number of equi-spaced *adjacent* channels increases. Obviously, by staggering in an irregular manner, the number of *on-channel* products can be modified or eliminated according to the spectrum space available. It is, however, difficult to show this effect in graphical form as the combinations are wide and varied.

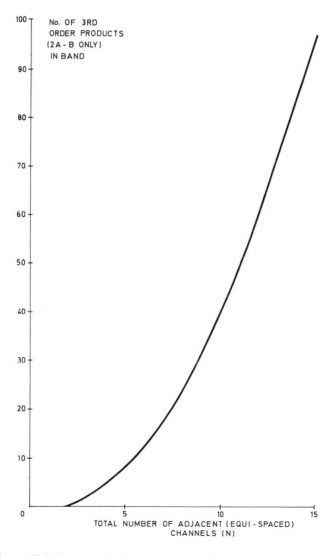

Fig. A6D.2 Number of Third Order (2A–B) Products Falling In-Band

Two simple formulae to enable the number of 2a–b products to be calculated for any number of adjacent equi-spaced channels (N) are

$$\frac{N}{2} \; (N-2) \quad \text{for even N}$$

and

$$\frac{N}{2} \; (N-2) \quad + \tfrac{1}{2} \text{ for odd N}$$

3.   Third order products of a + b – c configuration

As channel loading increases, the likelihood of more than two transmitters

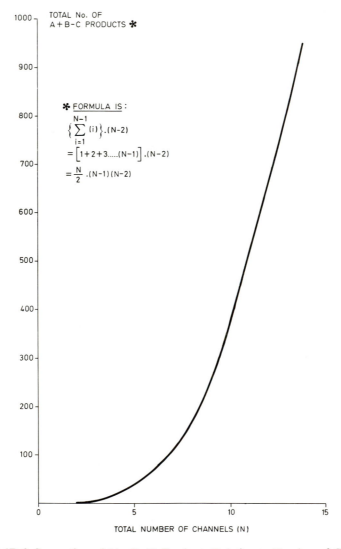

Fig. A6D.3 Generation of (A + B–C) Products Relative to Number of Channels

simultaneously radiating increases and the number of third order products based on the a + b – c configuration increases.

With this arrangement, the number of likely products increases very rapidly compared with the 2a–b case. It can be seen therefore that spectrum pollution can be extremely high around a communal site if care is not taken to minimise the generation of unwanted products both in the 2a–b and a + b – c categories.

3.1 Total number of products generated. Fig. A6D.3 attempts to show the magnitude of the problem if it is not restricted. The curve shows the rapid rate of increase in the total number of a + b – c products as the number of co-sited channels increases.

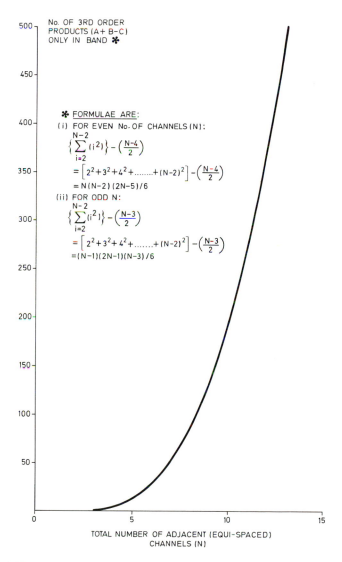

Fig. A6D.4 Number of Third Order (A + B–C) Products Falling In-Band

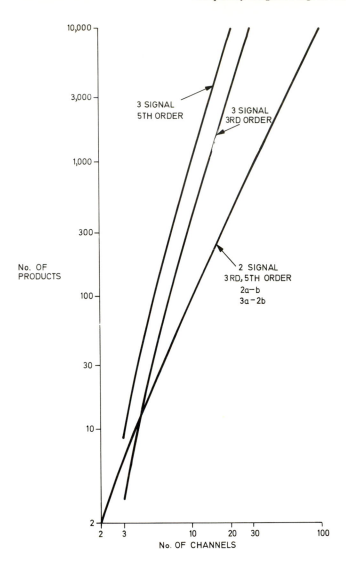

Fig. A6D.5 Comparison of Number of Products as Channels Increase in Number

3.2 a + b − c products generated in-band. As with the 2a−b products, it is the first essential to ensure that the systems on the site operate satisfactorily. The fraction of the total number of products shown in Fig. A6D.3 falling in-band is obviously important. Fig. A6D.4 plots the number of in-band products as a function of the number of equi-spaced adjacent channels located on the site.

As explained earlier, irregular spacing will remove some or all of the off-channel products but the considerable number of variations which can be used make this difficult to illustrate.

3.3 Fig. A6D.5 indicates the relationship between the various combinations.

4. Effect of reducing the number of operating channels per site

By reducing the number of channels per site, the reduction in intermodulation products (both 2a–b and a + b – c) can be dramatic.

For example, let us consider a site on which 10 channels are required. Suppose that, owing to a bandwidth switching limitation in the mobiles, these channels must be adjacent and equi-spaced.

Thus we have a total of 90 2a–b combinations which can be generated, of which 40 will lie within the spectrum used on that site.

Since irregular spacing is not possible in the example, extreme measures must be taken to minimise *radiated* products. However, nearby mobiles will receive high signal levels producing receiver non-linearity. Hence the above products can be generated within the mobile. There is no way of eliminating this effect and an area around the site will be virtually sterilised.

Now let us consider the a + b – c case. A total of 360 such combinations can be generated, of which 200 can lie within the spectrum used.

As with the 2a–b arrangement, only limited measures can be taken, and mobiles in the vicinity will be badly affected.

If, however, we now consider splitting the system and locating five of the channels on one site and five on a second site in another sector, we can immediately see an improvement.

In the 2a–b case, only 8 of the 20 possibilities fall in the part of the spectrum in use on the site.

Similarly with the a + b – c case, only 12 of the total of 30 combinations fall in-band.

Additionally, some channel staggering can now be used on each site, as shown in Appendix 8. This will further alleviate the effects to systems on the site although such staggering simply rearranges the combinations causing some to fall off critical channels.

5. Conclusions

This Annexe shows the degree to which the intermodulation problem on and around a communal site can develop if suitable measures are not adopted. It must be appreciated that the graphs show the *total* number of interference combinations possible; the *actual* number at any moment must depend largely upon the channel loading of the system. The chance of interference being caused on channels occupied in the neighbourhood must also be considered.

The proportion which the number of products present bears to the total number of possible products appears to follow the exponential curve normally associated with channel loading.

Appendix 7.   Intermodulation in VHF and UHF Radio Systems – Locating and Minimising the Effects

1.   General

Intermodulation and other spurious effects are described in detail in Appendices 6, 8, 9 and 10. Unfortunately, with the extensive use of communal sites each housing a considerable number of VHF and UHF systems, spurious effects for which there is often no immediate explanation have been known to arise, sometimes even without adding equipment on the site.

The following notes are intended as a guide to the isolation of spurious effects and to assist in determining the measures necessary to minimise the trouble.

2.   Basic Sources of Intermodulation

Intermodulation effects are produced when signals at two or more frequencies are applied to a non-linear device. Whether or not the resultant products will be troublesome must, of course, depend on many factors. For instance, the amplitude of the signals producing the effects may, in many cases, be controllable. As many of the "non-linear" devices do not show signs of gross non-linearity until overloaded, this aspect must be considered first.

The frequencies involved will determine where the spurious effects fall in the spectrum and so must be examined carefully.

In practice, intermodulation can arise in one of two locations, either (a) in the transmitters or their immediate environs or (b) in the receivers and their associated equipment. The first essential in tracing intermodulation is to determine whether it originates at the transmitter, receiver or is a combination of both.

3.   Separation of Intermodulation Effects

One of the most useful tools for the location of intermodulation is a variable RF attenuator. Connected between the antenna feeder and receiver input, it permits the incoming interference signal to be reduced in discrete steps and any overloading effect to be observed.

It can be seen that a genuine on-frequency signal arriving at the receiver input will be reduced in level at the same rate as the increase in attenuation; yet an intermodulation product caused by receiver overload will disappear immediately the signal causing the overload is attenuated below the overload level.

If, on the other hand, the intermodulation product drops at exactly the same rate as the introduced attenuation, then it is certain that the receiver is not the cause, and that the effect probably originates at the locality of the transmitter.

Certain points must, however, be considered when making the above measurement:

(*a*)   On a common transmitter and receiver site, direct breakthrough into later stages of the receiver may cause misleading results. Disconnection of the antenna will indicate whether or not this is occurring. If there is breakthrough into a later stage, then only by removing the receiver a short distance away can the true effect be determined.

(*b*)   A combination of transmitter and receiver intermodulation can be difficult to determine but, by careful adjustment of the antenna attenuator, a change in the attenuation slope will be observed. If this is so, then the cause(s) at the transmitting end of the system must be isolated by methodically switching off individual transmitters, changing antenna layouts or feeding certain transmitters into dummy loads. Only by systematically eliminating each possible source of intermodulation will it be possible to trace the origin.

(*c*)   A third cause of intermodulation generation is non-linearity in metalwork in the vicinity of the transmitter site.

To trace this source, a tunable receiver with its input connected to a small loop via a screened lead is recommended. By moving the loop around the suspected area of intermodulation generation, the exact source can often be pinpointed.

Such areas as mast joints, twisted guy wires, antenna support clamps, perimeter fences and corrugated iron roofs are all likely sources of signal rectification. Where the elements involved have physical proportions approximating to multiples of the affected wavelength, the radiation efficiency can be quite high.

4.   Reducing Intermodulation Interference

It is extremely difficult to suggest means of reducing intermodulation without knowing the specific intermodulation problem. The following notes are intended as a guide only and each problem must be examined in the light of existing circumstances.

Certain points are, however, fundamental. For instance, if intermodulation is found to be caused by receiver overloading, it is obvious that this can only stem from bad siting or bad frequency planning. Likewise, haphazard intermingling of transmitter antennae on a mast often produces unwanted spurious transmissions.

4.1   Receiver Overloading. Present day VHF and UHF communication receivers designed to the specifications existing throughout the world show practically identical overload characteristics. Blocking starts to occur with an equivalent on-channel signal of about $+80$ dB relative to 1 μV, whilst sufficient non-linearity will exist to produce initial intermodulation effects with a signal of about $+60$ dB relative to 1 μV.

Thus, where intermodulation effects are likely, the combined interfering signals must not be greater than 1 mV.

As we have seen already, protection against intermodulation caused by receiver overloading can be achieved by frequency spacing or physical distance, or by a combination of both, so as to reduce the on-channel signal to less than 1 mV.

As discussed earlier, with single frequency simplex only distance separation can help. Also, methods are suggested by which the minimum frequency spacing in co-sited duplex schemes can be calculated. Here distance cannot help much owing to the need for co-siting, although antenna spacing will provide some protection.

Certain advantages may be gained by special antenna systems. For example, an offending transmitter may be 0·4 km ($\frac{1}{4}$ mile) from the receiver and there may be no necessity for the receiver to receive signals from that direction. In such a case, a simple directional antenna having a front-to-back ratio of, say 15 dB would suffice to reduce the overloading signal to an acceptable level. Similarly, it may be possible to fit a directional antenna to the transmitter if it is not required to radiate in the direction of the receiver.

Reduced transmitter power can also assist and, in certain cases, may give quite acceptable coverage. Alternatively, the use of a directional antenna together with reduced power may be possible.

If the affected receiver is being used on a point-to-point link, it may be possible (if the wanted signal is high) to reduce the receiver effective sensitivity by inserting an antenna attenuator. This would improve the overloading level by the amount of attenuation and could, in a border-line case, eliminate the intermodulation. Again, a combination of increased antenna gains, narrow beam widths and reduced receiver sensitivity should be considered.

In cases of receiver overloading, where the effect is marginal and where the major portion of the available protection has been achieved by frequency spacing, additional selectivity in the receiver front end will undoubtedly assist. For instance, although every precaution may have been taken at a communal site, it is possible that a mobile transmitter on an entirely separate network may occasionally cause overloading when in the vicinity of the affected receiver. It may be that a frequency spacing of, say, 3 MHz already exists but, owing to the proximity of the mobile, overloading occasionally occurs. A filter in the receiver antenna lead will quite often provide another 20 dB, and this may be sufficient to reduce the overloading signal to an acceptable level. The filter should be of a type to peak at the wanted receiver frequency to ease tuning procedure and to provide protection against all unwanted signals.

In many parts of the world, intermodulation can often be attributed to VHF broadcasting and television stations. Located on extremely good sites and radiating high powers with wide bandwidth, they are very often the initial cause of receiver overloading. Other signal levels in the vicinity of the receiver, therefore, can be quite low for intermodulation to arise. The wide bandwidth of the offending broadcast transmitters makes the problem even more acute by extending the bands occupied by the intermodulation products.

Fortunately, the offending transmitter is easily identified by the programme content. Again, filters in the receiver input are usually effective with the frequency separation available.

Nevertheless, since both broadcast transmitter sites and communal VHF sites are usually high and on nearby hills (their coverage problems are usually similar) the problem is difficult to avoid.

4.2 Generation of Intermodulation on Transmitter Sites

4.2.1 *Generation in Transmitter Output Stages.* Fortunately, with the low power associated with VHF/UHF communication systems, the area over which intermodulation interference extends will be somewhat limited.

In most cases (except where the assembly of transmitters in a rack has not been undertaken with care) the intermodulation generation will be caused by coupling between antennae, either directly or via the feeder runs (or a combination of both).

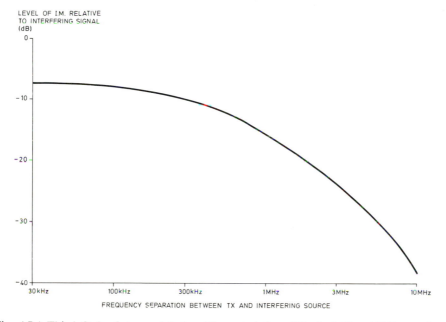

Fig. A7.1 Third Order Intermodulation Characteristics of Typical Class 'C' Transmitter Output Stage at 150 MHz

Fig. A7.1 shows the third order intermodulation level generated relative to frequency spacing and the level of interfering signal in a Class C stage. It can be seen that, at 3 MHz spacing, the intermodulation product could be 24 dB down relative to the level of the interfering transmission at the transmitter generating the product. If the offending transmitter is of 30 W output, and the antenna separation between transmitters achieves an attenuation of 40 dB, then the interfering signal fed into the second transmitter will be equivalent to 3 mW. This, in turn, will generate an intermodulation product of 3 mW − 24 dB = approximately 12 μW.

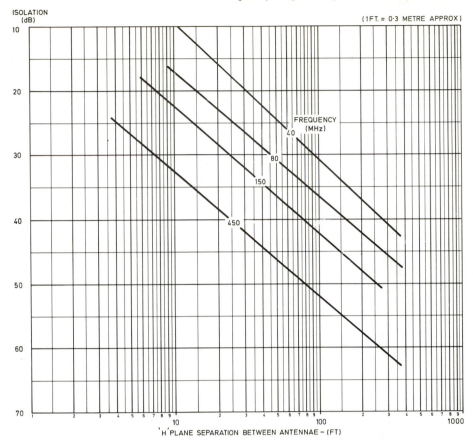

Fig. A7.2 Antenna Isolation as a Function of Horizontal Separation

Continuing the calculation, this level of signal could produce a 1 μV signal at a distance of up to 5 km (3 miles), assuming free space propagation at 150 MHz. With average terrain and obstructions, a similar signal would be received in a mobile unit at a distance of several hundred metres.

The intermodulation level generated in the output stages of transmitters on a communal site is best reduced by minimising the coupling rather than attempting to reduce the level at the mixing junction. The latter can only be achieved by making the output stage more linear and this, in turn, reduces efficiency. With an output stage operating in the most linear manner possible, the mixing of two signals in the amplifier would be some 50 to 55 dB down compared with the figure of 7 to 8 dB shown for adjacent channels in Fig. A7.1. However, this is only achieved with a considerable loss of efficiency and at a large increase in cost.

Therefore reduced coupling between transmitters is the better approach to the problem. A number of points can be outlined:

(a) Antenna layout should maximise the attenuation between the arrays in question. Maximum attenuation is always easier to obtain when antennae are in the vertical plane. The attenuations possible for horizontal and

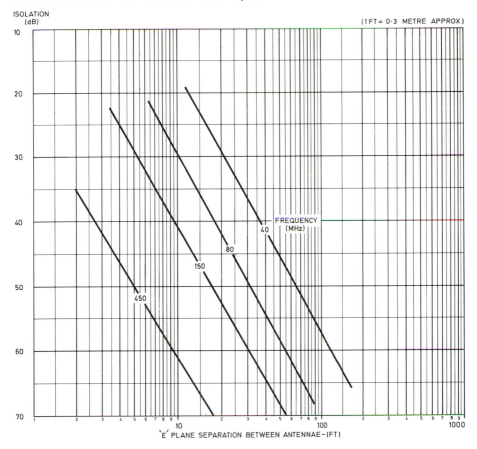

Fig. A7.3 Antenna Isolation as a Function of Vertical Separation

vertical separation are indicated in Figs. A7.2 and A7.3 respectively, based on vertical polarisation.

(*b*)   *Feeder Coupling*

Unless the coaxial feeder used employs completely continuous double screening, the leakage from a run of cable can be quite high. If two or more such cables are run in parallel there is obviously substantial coupling between the transmitters feeding the cables.

Thus, even if the antenna spacings are chosen with care, cable coupling can negate any advantage gained.

Where the cables are associated with critical transmitters, therefore, great care should be exercised in the layout of the cable run and separate ducts should be considered.

Let it next be assumed that both antennae and their feeders are installed in the best possible way and that interference still exists. What then is the next move?

Two possible arrangements can be considered:

(i)   If the transmitter frequencies are separated by over 2 MHz, then bandpass filters inserted between the transmitter output and the antenna feeder of one or both transmitters could eliminate or greatly reduce the inter-ference.

(ii) If the frequencies of the transmitters are close together, then the use of isolators in the transmitter outputs is indicated.

4.2.2 *Generation in On-Site Non-Linear Devices.* The level of intermodulation products generated by this mechanism cannot easily be calculated. Usually they fall far short of the products generated in transmitter Class C stages.

The level will obviously depend on the initial transmitter powers, the distance from the transmitters, non-linearity and, especially, the radiating efficiency of the device.

Thus, it is unlikely that interference will be a problem far beyond the site perimeter and, therefore, its effect (if the frequencies involved so allow) will normally be limited to on-site receivers.

The exception to this could well occur in the proximity of high power radio and TV stations, and radiation from non-linear devices subjected to such high levels could extend for some distance around the transmitter site. It should be mentioned here that the broadcasting mast itself could be the source of non-linearity, often in joints in the structure.

Reduction in the levels of interference caused by non-linear devices follows fairly conventional methods. Normally, non-linearity of this type is the result of contact between dissimilar metals or oxides formed at the junctions, aggravated by water or corrosive substances.

Such a non-linear element, in fields of sufficient magnitude generated by a number of transmitters, constitutes an ideal mixing medium. Combinations of frequencies of the originating transmitters are then produced.

Whilst the non-linear device can be cleaned and the effect eliminated, it may recur in areas of excessive corrosion such as salt-laden atmospheres. It is, therefore, often preferable to by-pass such joints by short and heavy gauge low impedance straps. These should be welded or soldered in position for best results.

Care should be taken to avoid resonance in these straps, as this will obviously worsen the effect.

Paint at fixing points can be another source of trouble. The earthing may then just be via a point contact which can corrode and become highly non-linear, even more so than large areas of dissimilar metals. The painted surface also introduces unwanted capacity effects, modifying the resonant characteristics of the area and often increasing the interference level.

By-passing the junction with a heavy strap is again the best way to minimise inter-ference generation.

It must be stressed, however, that in all the suggestions given above, the effect of intermodulation will, of course, only result in interference if frequency combinations are unfavourable. Therefore, the cure must always be preceded by the identification of the offending source(s).

Appendix 8. The Allocation of Frequencies in the VHF and UHF Bands A Method of Reducing Intermodulation Interference in Large Systems

1. General

The satisfactory operation of large integrated systems depends not only upon adequate engineering, but very largely upon the suitable choice of frequencies. This choice must, for example, involve the operation of the system as a whole, as well as problems associated with communal siting, multi-channel operation, single or double frequency operation and mobile density. It is appreciated that various P & T authorities ultimately decide the frequency allocation, but the official or semi-official user of the large system is often in a favourable position to obtain suitable channels.

Assuming that suitable blocks of channels can be made available, the following notes may be used as a guide in choosing suitable frequency combinations for minimum interference with the largest number of channels.

2. Intermodulation

The theory of intermodulation generation has been dealt with already in Appendices 6 and 7.

It should be emphasised that intermodulation is a fault condition, which must be minimised at the source, before considering further measures such as those which follow. Intermodulation, even if it does not affect the primary user, is often a source of trouble to other users in the vicinity.

Many of the problems and effects of intermodulation can be eliminated by two-frequency operation, using an adequate transmit-receive frequency spacing to avoid overloading of the receiving equipment. In large systems employing co-sited equipments, two-frequency operation based on two blocks of frequencies is essential. The following notes, outlining the optimisation of frequency planning, are based on this scheme. Under these conditions, co-siting can be achieved with minimum ill effect, the only critical factor being the effect on mobile receivers within a very short distance of the main base station.

3. Bandwidth Availability

In systems requiring a number of RF channels, one basic requirement is for the mobile units to switch over a band containing all the channels on which they must operate. The bandwidth must incur very little deterioration of performance and cause no damage.

Bandwidths of transmitters and receivers are fairly standard. To achieve adequate front-end performance, for instance, a receiver must have a selectivity at carrier frequency of better than a certain minimum. High selectivity will reduce the bandwidth over which an equipment can be switched, whilst low selectivity will reduce the ability of the equipment to discriminate against off-tune, interfering signals. In transmitters, such problems as distortion and excessive dissipation can arise as the circuits are off-tuned.

The normal bandwidth, accepted by many manufacturers as a compromise, approximates to $\pm 0.2\%$ of the nominal carrier frequency. This, for example, permits a spread of 600 kHz at 150 MHz. Certain equipments, however, have increased bandwidth design.

However, it must be borne in mind that the addition of antenna filters or duplexing equipment will reduce the total possible bandwidth, and this must be considered when determining the total available bandwidth.

4. Channel Capability

Having determined the total bandwidth over which either the transmitter or the receiver in a system will operate satisfactorily, two such blocks can be sub-divided into a number of operating channels each occupying a bandwidth determined by the local regulations. The blocks might, for example, be separated by at least 5·5 MHz (see Fig. A8.4) and one block would be allocated to the base station transmitters and the other to the base station receivers.

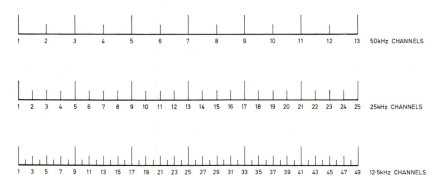

Fig. A8.1 Sub-division of 600 KHz Block

If one takes the example given in Section 3 above, of 600 kHz total bandwidth at 150 MHz, then each block can be sub-divided into 13 × 50 kHz channels, 25 × 25 kHz channels, or 49 × 12·5 kHz channels, etc.

Fig. A8.1 shows a block of 600 kHz divided in the manner outlined above; the channel numbering is used below.

5. Avoidance of On-channel Intermodulation

To avoid on-channel third order intermodulation products in a multi-frequency system, the allocation of channels must be staggered so that the spacing between *any two* channels in a block sequence is not repeated.

The following tables indicate channels clear of third order on-channel inter-modulation, with the final example showing third and fifth order avoidance.

Channels required	Channels available	Operating channels having no on-channel third order intermodulation interference
3	4	1, 2, 4
4	7	1, 2, 5, 7
5	12	1, 2, 5,10,12
6	18	1, 2, 5,11,13,18
7	26	1, 2, 5,11,19,24,26
8	35	1, 2, 5,10,16,23,33,35
9	46	1, 2, 5,14,25,31,39,41,46
10	62	1, 2, 8,12,17,40,48,57,60,62
11	78	1,17,18,24,28,43,56,64,73,76,78

Channels required	Channels available	Operating channels having no on-channel third and fifth order intermodulation interference
8	137	1, 2, 8,12,27,50,78,137

6. Choosing the Channel Sequence

The optimum sequence obviously depends on the number of channels finally required. If only a few channels are required then it is relatively simple to specify a sequence well within the total bandwidth of the equipment. For instance, if three 25 kHz channels are required, the above table shows that a total space of four channels only is required. This, occupying only 75 kHz, is well within the switching bandwidth of any VHF or UHF equipment, even when used with highly selective antenna filters or duplexers.

On the other hand, if the requirement is for as many channels as possible within the switching bandwidth, then the method must be modified accordingly.

Taking a further example, it can be seen that in a 600 kHz wide total band with 13 × 50 kHz available channels, five channels free of third order intermodulation can be allocated or four with freedom from both third and fifth order intermodulation. These are shown in Fig. A8.2 (i). Fig. A8.2 (ii) shows a similar treatment using 25 kHz channels, where totals of six actual channels free from third order inter-modulation and four with third and fifth order freedom are possible. Using 12·5 kHz spacing, the equivalent figures would be nine and five respectively (Fig. A8.2 (iii)).

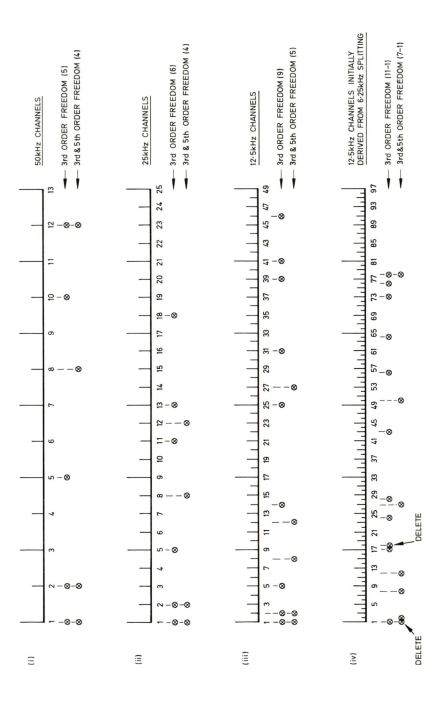

Fig. A8.2 Typical Choice of Channel Sequence

To increase the capacity still further requires sub-division of the basic channel bandwidths. For instance, by sub-dividing down to 6·25 kHz, a total of 97 channels becomes available. This permits a total of eleven channels free from third order intermodulation products or seven with third and fifth order clearance. However, 6·25 kHz channel spacing is out of the question on a single site. If 12·5 kHz equipment is in use, then by deleting one of the two channels having 6·25 kHz spacing, ten and six channels respectively based on 12·5 kHz are left. Fig. A8.2 (iv) refers to this technique and shows that it is achieved by deletion of channel 18 in the first instance and channel 1 in the second.

The sub-division can be made to suit individual requirements. For instance 25 kHz can be sub-divided to $8\frac{1}{3}$ kHz provided the two channels giving $8\frac{1}{3}$ kHz and $16\frac{2}{3}$ kHz spacing are deleted to leave a 25 kHz basic minimum.

7. Multi-site Operation

Where a number of multi-station sites are to be used, each serving a particular area but having possible overlap points where the perimeters merge, the method of allocation should be based on the above techniques, but could be a result of several sequences.

For instance, assume two blocks of 600 kHz (each of 25 × 25 kHz channels) are available and four working channels are required on each of six sites. Fig. A8.3 shows the process, based on giving third order intermodulation free allocation at each site and providing 25 kHz separation between any two frequencies in each block.

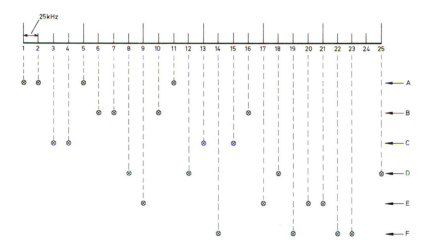

Fig. A8.3 Multi-Site Case

Step A   From the "6 from 18" sequence, the first four channels – 1, 2, 5 and 11 – are allocated.

Step B   Using the same sequence but staggering so as to start the sequence at channel 6, we allocate channels 6, 7, 10 and 16.

Step C   Using the same sequence again but starting at 3, deleting 5 and adding 13, we have 3, 4, 13 and 15.

Step D   Again using the same sequence but starting at 8, deleting 2 and adding 18, we have 8, 12, 18 and 25.

Step E   Reversing the sequence, deleting 11 and adding 13, we have channels 21, 20, 17 and 9.

Step F   Using the first four out of the "5 from 12" sequence, reversing the order and starting at 23 we have channels 23, 22, 19 and 14.

Thus, it can be seen that on each of six sites we have allocated four pairs of frequencies. Each site has third order on-channel intermodulation immunity and the same frequency has not been repeated throughout the six sites. This method is just an example of the flexibility of the system. By sub-dividing further, the combinations can be extended to points where even the most complex scheme can be considered.

If they are sufficiently distant, a further six sites can be equipped, using the same pattern, but staggered by 12·5 kHz. The distance may need to exceed 80 km, since this corresponds to an interfering signal level of 0·5 μV from a 50W transmitter at 150 MHz, based on free space loss.

The principle can also be extended to country-wide allocations, although in most countries the existing plan may be too far advanced for a change to take place.

8. Two-frequency Allocation – General

Certain points must be observed with two-frequency allocations, mostly associated with the avoidance of transmitter to receiver spacings which are related to the first intermediate frequency. Spacings such as ½IF, IF, 2IF are likely to lead to interference problems.

When two blocks are used, there is a possibility of random differences, caused directly or by off-channel intermodulation products. Under these circumstances it is as well to adjust the spacing between the *end* of the *first* block and the *start* of the *second* until it *just* exceeds ½IF; i.e. the spacing between the *start* of the *first* block and the *end* of the *second* block should be less than IF.

Fig. A8.4 shows an example with a first IF of 10·7 MHz. If the closest spacing is 5·5 MHz (just exceeding ½ of 10·7 MHz) then each block could extend to a width of approximately 2·5 MHz before the widest spacing reaches 10·5 MHz. Appendix 26 deals with this subject in greater detail.

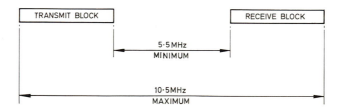

Fig. A8.4 Spacing of Blocks

Another point to observe is the generation of an intermodulation product a + b − c. This is discussed in detail in Appendix 6. However, some modification in the staggering as given in these notes and applied individually to the two blocks, can be used to remove the on-channel effect.

9. Conclusions

The method outlined in the foregoing paragraphs is intended to suggest how frequencies can be allocated so as to minimise the likelihood of third and fifth order on-channel intermodulation products. It is not intended as a method to eliminate intermodulation; this must be achieved at the source. It is merely a treatment designed to remove many of the remaining products off the wanted channels.

Certain limitations can exist, however, in practically any system, if a mobile operates within extremely short distances from the base station equipment. Under these circumstances such products as a + b − c can still be troublesome even with staggered spacings. However, for only a small percentage of time will there be coincidence between the mobile position and the use of critical frequencies.

The method makes possible a greater use of sites, particularly those forming part of a large integrated system. Obviously, its use is limited to systems where blocks of frequencies are allocated to the user; it is unlikely to be of great use where local authorities issue the individual frequencies.

Appendix 9.   Avoiding On-channel Intermodulation Products

1.   The Problem

After the necessary precautions have been taken to minimise the generation of inter-modulation products on communal or multi-user sites, one is often left with residual products. Their frequencies must be suitably arranged to avoid channels in use at that site or nearby.

The frequency plan involves staggering the channels by irregular amounts so that the resultant products fall off the channels in use. To achieve this result, the differences between any two channels must not be repeated throughout the sequence.

We are mainly concerned with third order products usually derived from the formula $2a - b$, but also including $a + b - c$ combinations where common transmit to receive spacings are used. Other third order products, such as $2a + b$, fall at frequencies widely separated from the band concerned.

In particularly bad cases, fifth order products are also of interest, $3a - 2b$ and variations of $2a - b - 2c$ being typical.

Appendices 6, 7, 8 and 10 deal in detail with the generation of intermodulation products, their identification and minimisation and contain simple combinations which are often adequate to clear the problem. In this Appendix, however, we have included the details of many combinations which can be used when planning frequencies for communal sites. The lists have been taken from a paper (Ref. 8) by R. Edwards, J. Durkin and D.H. Green, entitled "Selection of Intermodulation-free Frequencies for Multiple-channel Mobile Radio Systems". The author would like to express his thanks for permission to reproduce the lists.

CHANNEL DIFFERENCE SEQUENCES FREE FROM THIRD ORDER
INTERFERING INTERMODULATION PRODUCTS

Number of channels	Switching range	Difference sequences giving third order compatibility
3	6	2 3
	7	2 4
	8	2 5 3 4
	9	2 6 3 5
	10	2 7 3 6 4 5
	11	2 8 3 7 4 6
4	10	2 3 4 3 2 4 3 4 2
	11	2 5 3
	12	2 3 6 3 2 6 3 6 2 2 4 5 4 2 5 4 5 2
	13	2 3 7 3 2 7 3 7 2 2 6 4 3 4 5 4 3 5 4 5 3
	14	2 3 8 3 2 8 3 8 2 2 4 7 4 2 7 4 7 2 2 5 6 5 2 6 5 6 2 3 4 6 4 3 6 4 6 3
	15	2 3 9 3 2 9 3 9 2 2 4 8 4 2 8 4 8 2 2 7 5 3 7 4 3 5 6 5 3 6 5 6 3
5	15	2 4 3 5 2 4 5 3 4 2 5 3
	16	2 3 4 6 4 3 2 6 4 6 2 3 2 6 4 3 2 6 3 4 2 3 6 4
	17	3 2 4 7 4 2 3 7 2 4 7 3 4 2 7 3 4 7 2 3 2 3 7 4 2 5 3 6 2 5 6 3 5 2 6 3
	18	2 3 4 8 3 2 4 8 3 4 2 8 4 3 2 8 4 2 3 8 2 4 3 8 2 4 8 3 4 2 8 3 4 8 2 3 2 8 4 3 2 8 3 4 2 3 8 4 2 7 5 3 2 7 3 5 4 5 2 6 2 5 4 6 2 5 6 4 5 2 6 4 2 6 5 4 2 6 4 5
	19	2 4 9 3 4 2 9 3 4 9 2 3 2 9 4 3 2 9 3 4 2 3 9 4 2 5 8 3 5 2 8 3 2 3 6 7 3 2 6 7 6 2 3 7 2 6 3 7 2 6 7 3 6 2 7 3 6 7 2 3 2 3 7 6 2 4 5 7 5 7 2 4 2 4 7 5 3 4 5 6 4 3 5 6 5 3 4 6 3 5 4 6 3 5 6 4 5 3 6 4 5 6 3 4 3 4 6 5
	20	2 4 3 10 2 3 4 10 3 2 4 10 3 4 2 10 4 3 2 10 4 2 3 10 4 10 2 3 4 2 10 3 2 10 4 3 2 10 3 4 2 3 10 4 2 4 10 3 3 5 2 9 2 5 3 9 2 5 9 3 5 2 9 3 2 9 5 3 2 9 3 5 2 3 6 8 6 3 2 8 6 8 2 3 2 8 6 3 2 8 3 6 2 3 8 6 2 4 5 8 4 2 5 8 4 5 2 8 5 4 2 8 5 2 4 8 2 5 4 8 2 5 8 4 5 2 8 4 5 8 2 4 2 8 5 4 2 8 4 5 2 4 8 5 4 6 2 7 2 6 4 7 2 6 7 4 6 2 7 4 2 7 6 4 2 7 4 6 4 5 3 7 3 5 4 7 3 5 7 4 5 3 7 4 3 7 5 4 3 7 4 5
6	21	3 4 5 6 2
	22	2 4 5 3 7 3 5 7 4 2 5 3 7 4 2 3 7 5 4 2 5 3 7 2 4 3 5 7 2 4
	23	5 4 3 8 2 3 4 5 8 2 5 8 3 4 2 5 8 2 4 3 3 2 7 6 4 3 5 7 2 4
	23	5 4 3 8 2 3 4 5 8 2 5 8 3 4 2 5 8 2 4 3 3 2 7 6 4 6 4 7 2 3
	24	4 2 5 3 9 3 5 2 4 9 5 3 4 9 2 4 3 5 9 2 4 9 3 5 2 3 5 9 4 2 5 3 9 4 2 3 9 5 2 4 5 3 9 2 4 3 5 9 2 4 3 5 2 9 4 5 2 4 9 3 2 3 4 6 8 2 3 6 4 8 6 3 4 8 2 4 3 6 8 2 4 8 6 3 2 6 4 8 3 2 4 6 8 3 2 6 8 4 3 2 6 3 2 8 4 6 8 2 3 4 6 5 3 7 2 3 5 6 7 2 3 7 5 6 2 5 7 3 6 2 3 7 2 6 5
	25	2 4 3 5 10 2 4 5 3 10 3 10 4 5 2 5 10 3 4 2 3 5 10 4 2 5 3 10 4 2 3 10 5 4 2 3 10 5 2 4 5 3 10 2 4 3 5 10 2 4 3 10 2 5 4 5 10 2 4 3 5 4 2 10 3 5 2 4 10 3 3 2 6 4 9 3 4 6 2 9 6 4 3 9 2 3 4 6 9 2 4 3 9 6 2 3 4 9 6 2 6 4 9 3 2 4 6 9 3 2 6 2 3 9 4 3 2 6 9 4 3 2 9 6 4 6 2 9 3 4 6 2 9 4 3 6 4 9 2 3 3 2 4 7 8 4 2 3 7 8 4 2 7 3 8 3 7 8 4 2 7 3 8 4 2 3 8 7 2 4 7 8 3 2 4 3 2 8 7 4 7 8 4 2 3 7 4 8 2 3 7 2 4 8 3 3 6 5 2 8

CHANNEL DIFFERENCE SEQUENCES FREE FROM THIRD ORDER INTERFERING INTERMODULATION PRODUCTS
(continued)

Number of channels	Switching range	Difference sequences giving third order compatibility				
		3 6 5 8 2	3 6 8 5 2	6 3 8 5 2	6 3 8 2 5	3 6 8 2 5
		4 6 5 7 2	5 4 7 6 2	4 5 7 6 2		
6	26	2 4 3 5 11	5 3 4 2 11	4 5 3 11 2	5 4 3 11 2	
		5 3 4 11 2	3 5 4 11 2	3 4 5 11 2	4 3 4 11 2	
		4 11 3 5 2	5 11 3 4 2	3 5 11 4 2	5 3 11 4 2	
		5 3 11 2 4	3 5 11 2 4	3 5 2 11 4	5 11 2 4 3	
		4 3 2 6 10	2 6 3 4 10	6 2 3 4 10	4 3 6 2 10	
		6 3 4 10 2	4 3 6 10 2	4 3 10 6 2	3 4 10 6 2	
		4 10 3 6 2	4 10 6 3 2	6 10 4 3 2	3 6 2 10 4	
		6 3 2 10 4	6 2 3 10 4	3 2 6 10 4	6 10 2 3 4	
		6 2 10 3 4	6 2 10 4 3	2 4 7 3 9	2 3 7 4 9	
		4 9 7 3 2	7 4 9 3 2	4 7 9 3 2	3 7 9 4 2	
		7 3 9 4 2	3 9 7 4 2	7 3 9 2 4	3 7 9 2 4	
		7 3 2 9 4	3 9 5 6 2	6 9 3 5 2	3 9 6 5 2	
		3 9 6 2 5	6 9 2 5 3	3 8 5 7 2	5 8 3 7 2	
		3 8 2 7 5	2 5 4 6 8	2 5 6 4 8	6 5 4 8 2	
		4 5 6 8 2	6 8 4 5 2	4 6 8 5 2	6 4 8 5 2	
		4 8 6 5 2	6 8 2 5 4	6 5 2 8 4	3 5 4 6 7	
		5 3 6 4 7	4 6 3 5 7	4 5 6 7 3	5 7 4 6 3	
		6 7 4 5 3	4 6 7 5 3	6 4 7 5 3	4 7 6 3 5	
		6 7 3 5 4	6 3 5 7 4	6 2 5 9 3		
	27	4 2 5 3 12	3 5 2 4 12	3 12 4 5 2	4 12 3 5 2	
		5 12 3 4 2	3 5 12 4 2	5 3 12 4 2	3 12 5 4 2	
		3 12 5 2 4	5 3 12 2 4	3 5 12 2 4	3 5 2 12 4	
		3 12 2 5 4	5 12 2 4 3	5 4 2 12 3	5 2 4 12 3	
		2 3 4 6 11	2 6 4 3 11	4 3 11 6 2	3 4 11 6 2	
		3 11 4 6 2	6 4 11 3 2	4 6 11 3 2	6 11 4 3 2	
		3 2 11 6 4	3 11 2 6 4	6 11 2 3 4	6 2 11 3 4	
		6 2 11 4 3	6 4 11 2 3	3 2 4 7 10	3 2 7 4 10	
		7 4 10 3 2	4 7 10 3 2	7 10 3 2 4	7 2 3 10 4	
		3 2 7 10 4	3 2 10 7 4	7 10 4 2 3	7 4 10 2 3	
		4 2 8 3 9	3 4 8 2 9	3 4 8 9 2	4 3 9 8 2	
		3 4 9 8 2	3 8 9 4 2	8 3 9 4 2	3 9 8 2 4	
		3 2 9 8 4	8 2 9 3 4	8 2 9 4 3	8 4 9 2 3	
		8 2 4 9 3	3 6 2 5 10	3 6 5 2 10	5 6 3 10 2	
		6 5 3 10 2	6 3 5 10 2	3 6 5 10 2	3 5 6 10 2	
		5 3 6 10 2	5 10 3 6 2	3 6 10 5 2	6 3 10 5 2	
		6 3 10 2 5	3 6 10 2 5	3 6 2 10 5	7 3 5 9 2	
		3 7 5 9 2	3 5 7 9 2	2 3 6 7 8	3 2 7 6 8	
		2 3 7 6 8	3 6 7 8 2	3 8 6 7 2	6 8 3 7 2	
		6 8 7 3 2	7 8 6 3 2	7 2 3 8 6	3 2 7 8 6	
		7 8 2 3 6	4 6 2 5 9	5 9 4 6 2	4 6 9 5 2	
		6 4 9 5 2	6 4 9 2 5	4 6 9 2 5	4 6 2 9 5	
		2 4 5 7 8	7 5 4 2 8	5 7 4 2 8	7 4 5 8 2	
		4 7 5 8 2	4 5 7 8 2	5 4 7 8 2	7 8 5 4 2	
		4 2 8 7 5	7 8 2 4 5	7 5 8 2 4	3 4 5 6 8	
		5 6 3 4 8	6 4 5 8 3	5 4 6 8 3	4 8 5 6 3	
		6 5 8 4 3	5 6 8 4 3	6 8 5 4 3	6 8 3 4 5	
		5 3 7 9 2				
7	29	2 8 6 5 4 3				
	30	3 4 6 2 9 5	2 9 6 4 3 5	2 9 5 3 4 6	2 6 5 9 3 4	
		2 4 5 8 7 3				
	31	2 6 10 3 4 5	3 4 10 2 6 5	3 6 2 5 10 4	2 10 5 6 3 4	
		2 6 5 10 4 3	2 6 10 5 4 3	2 4 7 9 3 5	2 4 7 9 5 3	
		3 2 7 4 6 8	4 6 8 3 2 7	2 3 6 7 8 4	3 2 7 8 6 4	
		2 8 4 7 6 3				

CHANNEL DIFFERENCE SEQUENCES FREE FROM THIRD ORDER
INTERFERING INTERMODULATION PRODUCTS
(continued)

Number of channels	Switching range	Difference sequences giving third order compatibility			
7	32	2 6 4 3 11 5	3 11 4 6 2 5	3 11 2 6 4 5	2 6 11 3 4 5
		6 2 11 3 4 5	4 11 2 5 3 6	3 4 5 11 2 6	2 11 3 5 4 6
		2 5 11 3 6 4	3 11 5 2 6 4	2 5 3 6 11 4	2 6 11 5 4 3
		2 11 6 4 5 3	2 5 11 4 6 3	2 4 10 3 5 7	2 4 5 3 10 7
		3 5 10 4 2 7	3 5 10 2 4 7	3 5 10 2 7 4	3 5 10 7 2 4
		2 7 4 10 5 3	2 4 7 10 5 3	2 4 10 7 5 3	2 4 5 7 10 3
		4 3 9 2 8 5	3 9 8 5 2 4	2 8 5 9 3 4	2 9 5 8 4 3
		2 5 9 4 8 3	2 3 7 4 9 6	4 9 7 3 2 6	2 3 9 7 4 6
		4 9 6 2 3 7	3 2 6 4 9 7	2 3 9 6 4 7	2 6 9 3 7 4
		2 3 9 6 7 4	2 3 9 7 6 4	3 2 9 7 6 4	2 6 9 4 7 3
		2 8 7 5 6 3	2 6 4 5 11 3		
	33	2 6 3 4 12 5	4 3 6 2 12 5	4 6 3 12 2 5	3 6 4 12 2 5
		3 4 12 2 6 5	4 3 12 2 6 5	3 12 4 5 2 6	2 5 12 3 6 4
		3 12 6 2 5 4	2 6 5 12 3 4	2 6 5 12 4 3	2 12 5 6 4 3
		2 5 12 4 6 3	7 2 4 11 3 5	4 2 7 11 3 5	4 2 11 7 3 5
		4 2 11 3 7 5	4 2 11 3 5 7	4 2 11 5 3 7	3 5 11 4 2 7
		3 7 5 11 2 4	3 5 7 11 2 4	3 5 11 7 2 4	2 11 7 3 5 4
		3 10 8 4 2 5	3 4 2 10 8 5	3 4 2 10 5 8	3 10 5 2 4 8
		2 5 4 8 10 3	3 2 10 4 7 6	3 2 10 6 7 4	3 2 6 7 10 4
		2 3 6 7 10 4	3 2 7 6 10 4	3 2 9 8 4 6	3 2 9 6 4 8
		2 3 8 9 6 4	2 9 8 6 4 3	3 7 5 9 2 6	2 6 5 9 3 7
		2 6 7 3 9 5	2 6 5 9 7 3	2 8 4 5 6 7	4 8 2 7 6 5
	34	2 6 13 3 4 5	6 2 13 3 4 5	2 13 4 3 5 6	4 13 2 5 3 6
		3 4 5 13 2 6	2 5 3 6 13 4	2 13 6 5 3 4	2 6 13 5 4 3
		2 6 4 5 13 3	2 5 4 6 13 3	2 4 7 3 5 12	2 4 7 3 12 5
		7 3 12 2 4 5	3 7 12 2 4 5	7 4 2 12 3 5	4 7 2 12 3 5
		2 7 4 12 3 5	7 2 4 12 3 5	4 2 7 12 3 5	2 4 7 12 3 5
		2 4 12 7 3 5	2 4 5 12 3 7	4 2 12 5 3 7	2 4 12 5 3 7
		3 5 12 4 2 7	3 5 12 2 4 7	3 5 12 2 7 4	2 7 4 12 5 3
		2 4 7 12 5 3	4 2 7 12 5 3	2 4 5 12 7 3	2 8 5 4 3 11
		3 4 2 8 5 11	3 4 2 8 11 5	3 4 8 2 11 5	2 8 11 3 4 5
		8 2 11 3 4 5	2 4 3 11 8 5	2 4 3 11 5 8	3 4 5 11 2 8
		2 8 5 11 3 4	2 8 11 5 4 3	3 5 9 4 2 10	4 2 5 9 3 10
		4 2 9 5 3 10	3 9 10 4 2 5	9 3 10 4 2 5	2 9 10 4 3 5
		9 2 4 10 3 5	4 2 9 10 3 5	2 4 10 9 3 5	4 3 10 2 9 5
		4 2 10 3 5 9	4 2 5 10 3 9	2 4 10 5 3 9	3 10 9 5 2 4
		3 9 10 5 2 4	3 5 9 10 2 4	2 9 10 5 3 4	2 9 5 10 3 4
		2 10 5 9 4 3	2 5 10 4 9 3	4 6 7 2 3 11	7 2 3 11 4 6
		3 2 7 11 4 6	2 3 11 4 6 7	4 6 11 3 2 7	2 3 7 6 11 4
		3 11 7 2 6 4	3 2 7 11 6 4	2 3 11 7 6 4	2 3 4 8 6 10
		2 3 4 8 10 6	2 3 8 4 10 6	4 3 2 10 8 6	4 3 2 10 6 8
		2 3 6 10 4 8	2 3 6 10 8 4	2 3 10 6 8 4	2 10 6 8 3 4
		2 10 8 6 3 4	2 4 7 3 9 8	2 3 7 4 9 8	4 7 9 3 2 8
		3 9 7 4 2 8	3 7 9 2 4 8	2 4 9 7 3 8	2 3 9 4 7 8
		3 9 8 2 4 7	2 4 9 8 3 7	4 2 8 3 9 7	2 4 8 3 7 9
		3 2 8 4 7 9	4 7 8 2 3 9	2 3 7 8 9 4	3 2 9 7 8 4
		3 9 7 8 2 4	2 3 9 8 7 4	3 2 8 9 7 4	2 4 8 9 7 3
		2 8 7 4 9 3	2 4 7 8 9 3	3 5 6 7 2 10	2 7 5 3 10 6
		2 7 10 3 5 6	7 2 10 3 5 6	2 10 5 3 6 7	3 5 6 10 2 7
		2 6 3 10 5 7	2 6 3 10 7 5	3 10 6 2 7 5	2 10 7 6 3 5
		2 10 7 6 5 3	2 7 10 6 5 3	2 7 5 6 10 3	3 9 8 2 5 6
		2 5 6 8 9 3	2 6 9 4 7 5	2 6 4 7 9 5	2 9 5 7 6 4
		2 6 9 5 7 4	3 7 6 5 4 8	4 5 7 3 8 6	5 8 4 7 3 6
		5 4 8 3 7 6	4 8 5 6 3 7	3 8 6 4 5 7	4 6 3 8 7 5
		4 8 7 3 6 5	3 7 6 8 4 5	3 8 6 7 5 4	3 6 8 5 7 4
7	35	3 4 5 6 2 14	2 6 4 3 14 5	2 6 3 4 14 5	4 3 6 2 14 5

CHANNEL DIFFERENCE SEQUENCES FREE FROM THIRD ORDER INTERFERING INTERMODULATION PRODUCTS
(continued)

Number of channels	Switching range	Difference sequences giving third order compatibility			
		3 4 6 2 14 5	4 6 3 14 2 5	3 6 4 14 2 5	4 14 3 6 2 5
		2 6 14 3 4 5	6 2 14 3 4 5	2 14 6 4 3 5	2 14 3 4 6 5
		3 4 14 2 6 5	4 3 14 2 6 5	2 14 3 4 5 6	3 4 5 14 2 6
		2 14 5 3 4 6	2 14 3 5 4 6	2 5 14 3 6 4	3 6 2 5 14 4
		2 6 5 14 3 4	2 6 5 14 4 3	2 14 5 6 4 3	2 14 6 5 4 3
		2 6 14 5 4 3	2 14 6 4 5 3	2 5 14 4 6 3	2 4 5 3 7 13
		7 3 13 2 4 5	3 7 13 2 4 5	2 4 13 7 3 5	2 4 13 3 7 5
		2 4 13 3 5 7	2 4 5 13 3 7	2 4 13 5 3 7	3 5 4 2 13 7
		2 4 5 3 13 7	2 4 13 7 5 3	2 13 7 4 5 3	2 13 5 4 7 3
		2 4 5 13 7 3	2 4 13 5 7 3	5 8 3 4 2 12	3 12 8 2 4 5
		3 8 2 12 4 5	2 8 3 12 4 5	4 2 12 3 8 5	3 4 12 2 8 5
		3 4 2 12 8 5	2 4 3 12 8 5	3 4 2 12 5 8	2 4 3 12 5 8
		3 12 5 4 2 8	3 8 2 5 12 4	3 8 5 12 2 4	3 8 2 12 5 4
		2 8 3 12 5 4	2 12 5 8 3 4	2 8 5 12 4 3	2 4 5 8 12 3
		2 5 3 9 4 11	2 4 9 11 3 5	2 5 3 11 4 9	2 5 3 11 9 4
		2 5 3 9 11 4	2 11 9 5 3 4	2 4 9 11 5 3	2 3 5 4 12 7
		4 6 3 2 12 7	2 6 12 3 7 4	3 2 6 7 12 4	2 12 7 4 6 3
		2 6 12 4 7 3	2 7 4 6 12 3	2 3 4 6 8 11	6 8 4 3 2 11
		2 8 6 3 4 11	6 8 2 3 4 11	3 2 11 8 4 6	2 3 11 8 4 6
		2 8 11 4 3 6	3 2 11 4 8 6	3 4 11 2 8 6	4 3 2 11 8 6
		2 3 4 11 8 6	2 3 4 11 6 8	4 3 2 11 6 8	3 2 11 4 6 8
		2 3 4 6 11 8	2 3 11 6 4 8	3 2 11 6 4 8	3 2 11 6 8 4
		3 6 11 2 8 4	3 2 8 6 11 4	3 2 11 8 6 4	2 8 11 6 3 4
		2 8 6 11 4 3	2 11 6 8 4 3	2 8 4 11 6 3	3 4 9 6 2 10
		3 2 6 9 4 10	3 4 9 10 2 6	3 2 6 10 4 9	3 2 6 10 9 4
		3 2 6 9 10 4	3 2 4 7 8 10	3 8 7 2 4 10	7 8 3 2 4 10
		4 10 8 3 2 7	3 8 10 4 2 7	2 3 10 8 4 7	2 4 10 3 8 7
		3 2 4 10 8 7	3 2 4 10 7 8	4 10 7 2 3 8	3 2 7 4 10 8
		2 3 10 7 4 8	3 2 4 7 10 8	2 7 10 3 8 4	3 8 7 2 10 4
		3 2 7 8 10 4	2 3 8 7 10 4	3 8 10 7 2 4	3 8 7 10 2 4
		2 7 10 4 8 3	2 4 10 7 8 3	2 4 8 7 10 3	5 7 3 11 2 6
		3 7 5 11 2 6	5 11 3 7 2 6	2 11 7 5 3 6	2 11 6 3 5 7
		2 6 11 3 7 5	2 11 6 3 7 5	3 7 2 6 11 5	2 7 6 11 3 5
		2 7 6 11 5 3	2 6 11 5 7 3	2 6 7 5 11 3	3 6 5 2 10 8
		3 6 10 2 5 8	2 5 8 10 6 3	2 10 8 5 6 3	2 9 3 7 8 5
		2 9 5 8 7 3	2 9 8 5 7 3	5 4 7 6 2 10	2 6 7 5 4 10
		7 4 5 10 2 6	5 4 7 10 2 6	2 10 4 5 6 7	5 4 10 2 6 7
		4 10 7 2 6 5	4 10 2 7 6 5	2 6 7 10 4 5	2 10 7 6 5 4
		2 6 5 7 10 4	4 6 8 5 2 9	6 4 8 5 2 9	2 5 9 8 4 6
		5 2 9 8 4 6	2 9 5 8 4 6	5 2 9 4 8 6	5 2 9 4 6 8
		5 2 9 6 4 8	2 5 9 6 4 8	4 8 6 9 2 5	4 6 8 9 2 5
		2 8 4 9 6 5	2 5 6 9 8 4	4 9 7 3 5 6	3 5 7 9 4 6
		4 6 5 3 9 7	3 5 6 4 9 7		
8	40	3 5 6 4 12 7 2	5 3 12 6 4 7 2	3 7 12 4 5 6 2	
		3 7 4 5 12 6 2	5 4 7 3 12 6 2	4 5 12 2 6 7 3	
		4 5 6 2 12 7 3	4 8 2 5 11 6 3	4 3 9 10 5 6 2	
		9 3 4 10 5 6 2	5 7 8 3 10 4 2	6 9 7 4 8 2 3	
	41	3 5 6 4 13 7 2	4 6 5 3 13 7 2	4 6 13 3 5 7 2	
		6 2 13 5 4 7 3	5 6 3 4 12 8 2	4 3 6 5 12 8 2	
		5 2 12 8 3 6 4	8 3 6 4 12 2 5	5 4 8 7 3 11 2	
		3 7 11 5 8 4 2	2 4 7 9 3 5 10	3 5 10 9 7 4 2	
		7 4 2 10 9 5 3	2 3 6 7 8 4 10	6 8 4 7 10 3 2	
		8 4 10 7 6 3 2	4 7 6 8 10 2 3		

CHANNEL DIFFERENCE SEQUENCES FREE FROM FIFTH AND THIRD ORDER INTERFERING INTERMODULATION PRODUCTS

Number of channels	Switching range	Difference sequences giving fifth order compatibility
4	13	3 7 2 3 4 5
	14	3 8 2 2 5 6 5 2 6 5 6 2
	15	2 3 9 3 2 9 3 9 2
	16	3 10 2 2 5 8 2 6 7 6 2 7 6 7 2 3 5 7 4 5 6
	17	2 3 11 3 2 11 3 11 2 5 9 2 3 4 9 4 3 9 4 9 3 4 5 7 5 4 7 5 7 4
	18	2 3 12 3 2 12 3 12 2 2 6 9 6 2 9 6 9 2 2 7 8 7 2 8 7 8 2 4 10 3 3 5 9 5 3 9 5 9 3 4 6 7 6 4 7 6 7 4
	19	2 3 13 3 2 13 3 13 2 2 5 11 5 2 11 5 11 2 4 11 3 3 7 8 7 3 8 7 8 3 5 6 7
	20	2 3 14 3 2 14 3 14 2 5 12 2 2 6 11 6 2 11 6 11 2 2 7 10 7 2 10 7 10 2 2 8 9 8 2 9 8 9 2 3 4 12 4 3 12 4 12 3 5 11 3 3 7 9 7 3 9 7 9 3 4 6 9 6 4 9 6 9 4 5 6 8 6 5 8 6 8 5
	21	2 3 15 3 2 15 3 15 2 2 5 13 5 2 13 5 13 2 7 11 2 3 4 13 4 3 13 4 13 3 3 5 12 5 3 12 5 12 3 3 8 9 8 3 9 8 9 3 4 5 11 5 4 11 5 11 4 4 7 9 7 4 9 7 9 4 5 6 9 6 5 9 6 9 5 5 7 8 7 5 8 7 8 5
5	26	5 13 3 4 5 13 4 3
	27	2 6 11 7 3 9 10 4
	28	2 8 14 3 2 3 9 13 2 12 8 5 2 12 5 8 2 9 6 10 2 9 10 6 5 15 3 4 3 4 15 5 4 9 3 11 3 11 9 4
	29	2 16 7 3 2 16 3 7 2 15 8 3 2 15 3 8 3 2 11 12 2 3 12 11 2 6 15 5 6 2 15 5 6 15 2 5 2 5 15 6 2 9 12 5 5 16 3 4 3 16 5 4 3 16 4 5 3 4 16 5 3 10 11 4 3 13 7 5 3 13 5 7 3 9 11 5 8 5 6 9 5 8 9 6
	30	2 7 17 3 7 2 17 3 3 2 9 15 9 15 2 3 6 16 2 5 2 5 16 6 2 13 9 5 2 11 10 6 2 11 6 10 2 7 8 12 2 7 12 8 5 17 3 4 3 17 5 4 3 17 4 5 3 4 17 5 4 10 3 12 3 12 10 4 3 12 4 10 4 14 6 5 4 14 5 6
	31	2 9 16 3 2 3 16 9 3 2 12 13 2 3 13 12 5 2 6 17 6 2 5 17 2 6 17 5 6 2 17 5 6 17 2 5 2 5 17 6 2 11 5 12 2 11 12 5 11 2 12 5 2 6 13 9 7 2 8 13 2 7 13 8 7 10 2 11 2 10 7 11 2 10 11 7 2 11 10 7 5 18 3 4 3 4 18 5 3 11 12 4 5 9 3 13 3 13 9 5 3 10 9 8 4 7 10 9
	32	2 7 19 3 7 2 19 3 2 19 7 3 2 19 3 7 2 18 8 3 2 18 3 8 9 17 2 3 2 17 9 3 2 17 3 9 2 3 17 9 2 10 16 3 10 2 16 3 2 3 11 15 11 15 2 3 2 3 15 11 2 6 18 5 6 2 18 5 6 18 2 5 2 18 6 5 2 18 5 6 2 5 18 6 5 9 2 15 2 15 9 5 2 15 5 9 2 16 7 6 2 16 6 7 6 9 2 14 2 14 9 6 2 14 6 9 2 13 10 6 2 13 6 10 8 9 2 12 2 12 9 8 2 12 8 9 5 19 3 4 3 19 5 4 3 19 4 5 3 4 19 5 3 4 9 15 3 4 15 9 4 11 3 13 3 13 11 4 3 13 4 11 5 11 3 12 3 12 11 5 3 12 5 11 3 8 7 13 3 8 13 7 6 16 4 5 4 5 16 6 4 5 7 15 7 15 4 5 4 5 15 7 6 7 8 10 6 7 10 8

2. Use of the Lists

The tables clearly indicate the basis on which they are normally used. The second column shows the channels over which the mobile equipment must be able to switch. Most equipments will switch, without exceeding a predetermined limit of degradation, over a band of $\pm 0.2\%$ of the centre frequency. This means that at, say,

150 MHz, provided the equipment is aligned at 150 MHz, the performance from 149·7 to 150·3 MHz will be within specification.

It can be seen that this frequency spread includes a total of 24 channels each 25 kHz wide, 48 channels 12½ kHz etc., in the band around 150 MHz.

As the nominal frequency is reduced, the actual frequency spread is also reduced, so that, for instance, only half the number of channels can be based on a 75 MHz centre frequency.

Some later designs of mobile equipment have a greater frequency spread, up to ±0·5% being available on certain units. Conversely, where greater front-end selectivity may be needed, as in fixed station applications, then the switching bandwidth must be restricted and ±0·1% is typical in these cases.

Here it should be pointed out that, in many cases, only the mobile units require the full switching range in order to switch to various base stations throughout a system. A base station unit may often be tuned to a single frequency, so the switching bandwidth restriction here may not be so serious.

Furthermore the communal site may contain systems with several users but individual users may require only one or two channels. Thus the distribution of frequencies on a communal site can often include the complete available frequency block(s), but individual systems would normally only switch over a small sequence within this block.

Referring again to the lists, the first column shows the number of on-channel interference free products possible with the channel switching range given in column 2. The main list shows the intervals necessary between successive channels to achieve the on-channel freedom specified.

As an example, let us take the first four-channel combination giving four free channels out of the ten available. This combination 2, 3, 4 indicates on-channel freedom if Channels 1, 3, 6 and 10 are used.

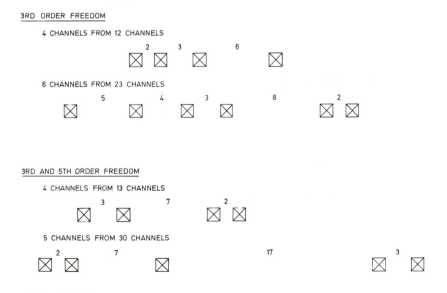

Fig. A9.1 Examples of On-channel Intermodulation – Free Channels
(Transmit Block Only Shown)

Fig. A9.1 shows several examples taken from the lists whilst Fig. A9.2 shows a further example hypothetically divided amongst several users on the same site.

The listed combinations can, if required, be used in the reverse order provided the sequence is retained. Some of the examples given in Section 12 in the Guide adopt this method to achieve the maximum utilisation of a given block. In fact, if reverse sequences are added to the lists attached to this Appendix, the number of combinations almost doubles.

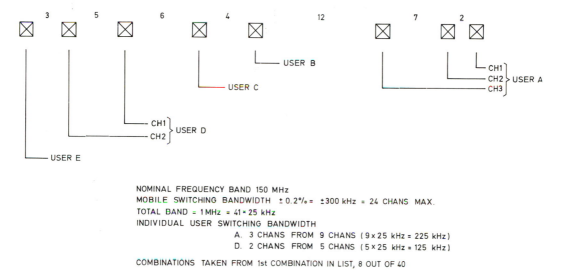

NOMINAL FREQUENCY BAND 150 MHz
MOBILE SWITCHING BANDWIDTH ± 0.2% = ±300 kHz = 24 CHANS MAX.
TOTAL BAND = 1 MHz = 41 × 25 kHz
INDIVIDUAL USER SWITCHING BANDWIDTH
        A. 3 CHANS FROM 9 CHANS (9 × 25 kHz = 225 kHz)
        D. 2 CHANS FROM 5 CHANS (5 × 25 kHz = 125 kHz)

COMBINATIONS TAKEN FROM 1st COMBINATION IN LIST, 8 OUT OF 40

Fig. A9.2 Typical Arrangement of Communal Site for Five Users Free of Third Order Intermodulation (Transmit Block Only Shown)

Fig. A9.3 shows a typical plot of one section of the list, dealing with four combinations from up to fifteen choices based on third order on-channel intermodulation freedom. All the lists can be plotted in this manner but the result, although laborious, may be clearer than the list of figures. It is suggested that, when the number of available channels in a block is known, the combinations are shown for that number as in Fig. A9.3.

Fig. A9.4 shows a few of the seven from 34 on-channel third order freedom possibilities. Again the listed combinations can be increased by a factor of two by reversing any of the sequences, giving a grand total of 264 combinations in this case.

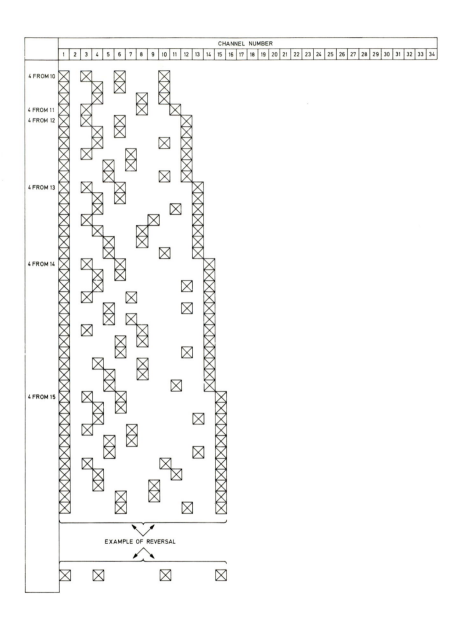

Fig. A9.3 Typical Plot from Tables – Third Order Freedom

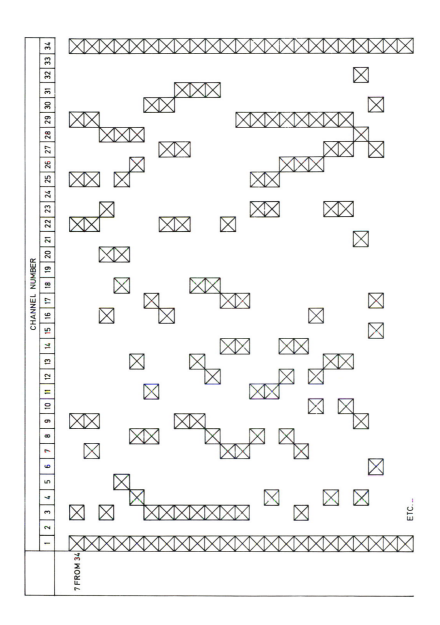

Fig. A9.4 Some of the Seven from 34 Combinations

Appendix 10. The Use of Circulators and Isolators to Minimise Transmitter Intermodulation

1. General

One of the most prolific sources of intermodulation on communal sites is the Class C output stage of conventional transmitters. The non-linearity characteristics of this mode of amplification are well known (see Appendices 6 and 7) and, at the relatively high levels of power involved, quite a small unwanted signal fed into such a stage from a second source can cause unwanted mixing, resulting in the generation of spurious signals on related frequencies.

On a communal site, this can be provided by coupling between antennae, feeders (particularly if excessive standing wave ratios exist) and between equipments. Unwanted coupling can be minimised by good engineering, but often sufficient remains to allow some unwanted products to be generated.

Where such products fall on channels unused in the vicinity, then often, on the grounds of economy, the shortcoming is tolerated. However, when such products fall on local channels, in use or projected, steps must be taken to eliminate the trouble.

One method is to ensure that the frequency plan prevents the products falling on wanted channels (see Appendices 6, 8 and 9) and this should be a standard procedure to adopt for the site frequencies. However, radiation of the unwanted off-channel products can extend over considerable areas. To minimise this short-coming, other measures are employed, including coaxial filters, circulators and isolators.

2. The need for Circulators or Isolators

Where unwanted products fall in a part of the frequency spectrum removed by megahertz rather than kilohertz from the wanted frequencies, coaxial filters are often used. Such filters are of two types, bandpass and band-reject. In the former case they are resonant at the wanted frequency and reject all other frequencies, by an amount dependent mainly upon the separation. On the other hand, the band-reject filter is normally tuned to the unwanted signal so as to reject it whilst allowing all others to pass.

The actual type and its construction depend upon the circumstances and, under the correct conditions, the resulting rejection can be high, with a relatively low attenuation of the wanted signal.

For confined spaces, the coaxial filter is often replaced by a lumped circuit filter using conventional tuned circuits, but the performance of such devices is inevitably inferior to the correctly designed coaxial cavity filter.

In either type, as the frequency spacing between the wanted and unwanted signals is reduced, the filter's rejection efficiency worsens. Its insertion loss at the wanted frequency can also increase if an attempt is made to improve the rejection by series connection of several filters or reduced coupling.

In these circumstances, circulators can be considered. With a terminated circulator (also known as an isolator), transmitter coupling can be reduced even to adjacent channels. The result would not be possible with transmitter filters although low power crystal filters are available for receivers to attenuate adjacent channels above 150 MHz.

3. Circulator/Isolator Operation

A circulator is a passive non-reciprocal device with three or more ports (inputs or outputs). Usually, it contains a core of ferrite material in which energy introduced into one port is transferred to an adjacent port, the other port(s) being isolated.

Although circulators can be made with any number of ports, 3 or 4 are most common. Fig. A10.1 shows these.

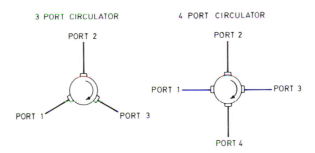

Fig. A10.1 Three- and Four-Port Circulators

Energy entering into port 1 emerges from port 2, energy entering into port 2 emerges from port 3 and so on in cyclic order. In this direction of circulation, an ideal circulator would have no losses, but in practice a slight loss (about 0·5 dB) occurs.

In an ideal circulator, no energy would flow in the direction opposite to the circulation direction. Again, in practice this isolation is of the order of 30 dB, or higher in some narrow band devices.

This non-reciprocal behaviour is the result of gyromagnetic effects in the ferrite when this is biased with a high frequency magnetic field.

Isolators (see Fig. A10.2) are basically 3 port circulators with one port terminated by its characteristic impedance. Thus, in the forward direction, energy can pass between ports 1 and 2, whilst energy arriving at port 2 is prevented by the circulator action from emerging from port 1. Instead it passes in the direction of circulation to port 3 where it emerges to be absorbed by the terminating load. The accuracy of this termination is of great importance. An imperfect standing wave ratio will cause energy to be reflected. In turn, this will re-enter the isolator and continue on its circulatory path to emerge from port 1. Correct terminations are, of course, important at all ports for optimum performance.

Since both circulators and isolators employ quite strong internal magnetic fields, carefully adjusted for optimum performance, external fields should be avoided.

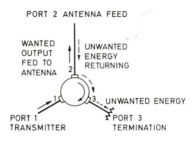

Fig. A10.2 Isolator

4.   Definition of terms

Some of the terms used to specify isolators are defined as follows:

(*a*)   *Frequency range:*
   The range within which the isolator meets the guaranteed specification. Outside this range the electrical properties deteriorate rapidly although no damage to the circulator will result.

(*b*)   *Isolation:*
   The attenuation through the device in the reverse direction, measured with matched source and load, and usually expressed in dB.

(*c*)   *Insertion loss:*
   The attenuation of the wanted signal resulting from the insertion of an isolator into a transmission system, expressed in dB. It is the ratio of the power delivered to a matched load before insertion of the isolator to the power delivered to that load after insertion of the device.

(*d*)   *VSWR:*
   The ratio of the maximum to the minimum voltages along the feed line.

(*e*)   *Nominal power:*
   The maximum power that may be passed through the isolator in the forward direction into a terminated port at an agreed VSWR, usually 2:1.

(*f*)   *Temperature range:*
   The ambient temperature range within which the isolator functions to specification. The isolator will continue to function outside this range but

some of its characteristics may change. Storage temperatures are usually well in excess of operating temperatures.

5. Method of use

Fig. A10.3 (i) shows two adjacent transmitters, operating into separate antennae and on nearby channels (i.e. within 1%). With antenna systems mounted on a common structure, provided care is taken with feeder runs, polarisation and spacing, there may be about 30 dB isolation between antennae. Any greater coupling would cause very high levels of re-radiated products.

Accepting the figure of 30 dB and adding a conversion figure of 7 dB (see Appendix 7, Fig. A7.1) gives a total of 37 dB. When related to a 25 watt transmitter level, this gives third order products at a level of approximately 5 mW, fifth order products being somewhat less. Based on average propagation, this level of radiation could produce about 1 μV into a mobile receiver at a distance of 1·5 miles (2 km).

With optimum frequency planning, this level may not be serious. However, in the event of unfavourable planning, this may be considered excessive, and isolators can provide a solution.

Fig. A10.3 (ii) shows an isolator inserted in each transmitter output feed. Whilst the outgoing power remains relatively unaffected (less than 1 dB down), the power arriving back at either transmitter owing to antenna coupling will be reduced by the isolation (about 30 dB).

The 5 mW now becomes 5 μW and the interference range is thus reduced to under 100 metres for the same level of interference as before.

We thus have a simple method of reducing transmitter intermodulation when the channel spacings are too small for the efficient use of coaxial filters. The insertion losses encountered in these devices are quite small, permitting series connection of two or more if additional isolation is required.

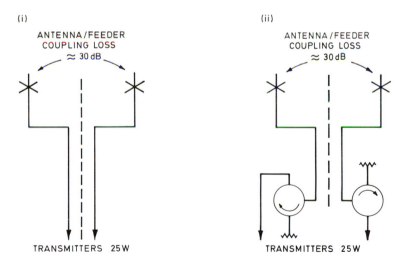

Fig. A10.3 Improvement by the Addition of Isolators

6. Pitfalls

Although isolators provide an apparently simple solution to transmitter inter-modulation, there are a number of avoidable pitfalls, which must not be ignored.

6.1　Terminating load power rating. Examining Fig. A10.4, it might appear that the power at the terminating resistance, R2 for example, would be of the order of 30 dB below the output power of Tx 1, and vice versa. It would seem that, short of the two antennae becoming enmeshed by, say, wind damage, this level would be unlikely to increase.

This assumption is undoubtedly partially correct. However, one advantage of an isolator is to prevent changes in the antenna from upsetting its own transmitter. So, if the antenna becomes open circuited, short circuited or affected by, say, snow or ice loading, then the isolator will prevent transmitter damage caused by mismatch.

In doing so, it can be seen that the reflected power must be dissipated in some way and this occurs in the terminating load, R1 or R2. Thus the rating of this load must not be appreciably lower than the transmitter output. A few watts may not matter but a level such as 100 watts in a 20 watt load would be unacceptable. (*Both* loads could be affected if the two antennae became enmeshed.) The overload could damage the termination, and, in extreme cases (exacerbated by high ambient temperature), it may be destroyed. If so, the isolation would cease to exist and transmitter damage could result.

Thus it is essential to specify a termination load power not appreciably less than the transmitter output.

In the Microwave Associates' leaflet (Ref. 15) affecting the warranty of their circulators and isolators, the following restrictions appear:

'Note. When the 20-watt field-replaceable termination is used, minimum dissipation occurs and no long-term protection is achieved. Reflected power in excess of 20 watts will occur if the antenna or antenna cable is removed

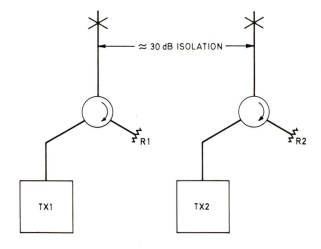

Fig. A10.4 Termination Power Rating

during transmitter operation and may occur during extended periods of antenna ice loading or from rain penetration of antenna cable or connectors. Units failing due to termination burn-out are not covered under the warranty.'

This statement obviously refers to transmitters exceeding 20 watts and therefore isolators connected to such transmitters should follow the recommendations of Microwave Associates. Below 20 watts, the field-replaceable termination should be adequate to cope with antenna disconnection, etc.

If more than one isolator is connected in series, isolators nearest the transmitter can be terminated with relatively small loads. Fig. A10.5 shows the power levels in the fault condition and also in the normal case.

Microwave Associates and, presumably, other manufacturers claim that, for maximum practical reduction in transmitter-generated intermodulation, up to three series isolators may be connected in series.

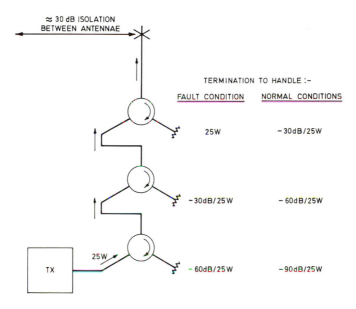

Fig. A10.5 Power Levels in Multiple Isolator Configuration

6.2 Unit "Hop-over". One should check that unwanted coupling does, indeed, originate from antenna coupling. Another possible cause is coupling or hop-over between equipments or cabinet wiring. Any device placed in the transmitter feeder run will only reduce the levels of signals arriving from outside. Equipment coupling will naturally cancel the benefit of isolators. With units in the same cabinet, tests involving isolators in antenna feeds may reveal an excessive irreducible minimum. Then it is as well to check isolation with greater spacing or screening covers between units.

Common earthing of a number of feeder lines within a cabinet can also introduce unwanted coupling and should be checked in the event of a large irreducible minimum. Similar trouble can be caused when coaxial filters are used, constituting one of the many problems in achieving the design attenuation.

A poor quality feeder within the cabinet, or a badly made coaxial joint, is another source of hop-over.

6.3   Power rating of isolators. The power rating of a circulator or isolator should be observed for proper performance of the system over the whole frequency band. Excessive power causes the magnetic material in the isolator to approach saturation and the resultant non-linearity increases harmonic distortion in the transmitter output. The necessary reduction in power for satisfactory performance may vary between manufacturers. To ensure relative freedom from non-linearity, isolators should, wherever possible, be used at about 50% of the maximum rating.

6.4   Magnetic effects. As mentioned in 3 above, magnetic fields can affect the units. Mounting the units on the wall of the room at a point of feeder entry may be preferable to housing them within the cabinet. Certainly, components such as loud-speakers and transformers should be kept well away.

6.5   Bandwidth. The performance of circulators or isolators falls off beyond a certain percentage of the nominal frequency, although damage will not result. Total bandwidths of 3 to 4% of the nominal frequency are usual and at these limits the isolation usually drops by 15 to 20 dB. Thus, a 150 MHz unit could be used from 147–153 MHz but, for maximum isolation, a more realistic band might be 148·5 to 151·5 MHz ($\pm$1%) giving a band edge loss of isolation of not more than 4 to 5 dB. Fig. A10.6 shows the performance of typical isolators.

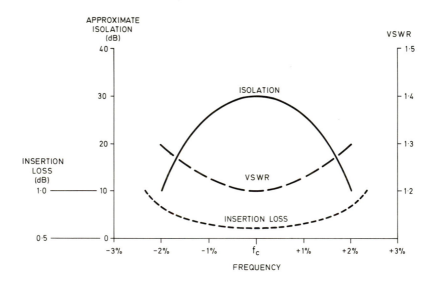

Fig. A10.6 Typical Isolator Performance

At the band extremes, the performance is similar to a standard type of coaxial filter. So beyond, say, $\pm$2%, a coaxial filter would undoubtedly score. Two or more isolators can be used in series, each providing isolation in discrete but adjacent $\pm$1% sections of the band to provide a greater overall bandwidth. However, the insertion loss of each isolator would be greater, so the total may be unacceptable.

7. Combining isolators and filters

On communal sites, several frequencies may be instrumental in causing an inter-modulation problem. Furthermore, certain of these frequencies may be separated from the others by several MHz. Whilst isolators will perform satisfactorily in the closely spaced channels, the separated channels will not benefit to the same degree.

Under these conditions it may be necessary to consider a combination of isolators and filters.

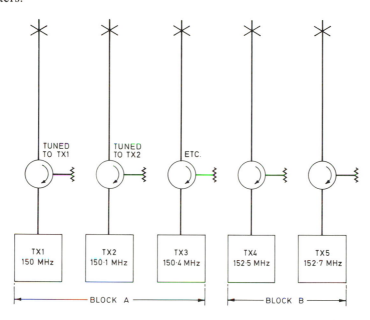

Fig. A10.7 Isolation With Wide Frequency Spacing

Fig. A10.7 illustrates such an example. If we assume antenna spacing to provide about 30 dB isolation, then the isolation between pairs of transmitters will be 30 dB plus that provided by the isolator. For pairs of transmitters (such as Tx 1 and Tx 3) in the same block, the isolators will provide about 30 dB, giving a total isolation of 60 dB. However, reference to Fig. A10.6 shows that the isolator performance will drop to about 10 dB between blocks (i.e. about 1·8% off-tune in this case). If the blocks were even further separated, the performance would be worse. Indeed, beyond about 4 MHz apart, the isolation between (for example) Tx 1 and Tx 4 would be merely 30 dB, derived from antenna separation.

Fig. A10.8 shows a solution using bandpass filters, presumed to give 30 dB attenuation of the other block. The isolation between transmitter pairs in the same block will be 60 dB, as before (filter attenuation = 0 dB). Between pairs of blocks, the filter will contribute a further 30 dB, giving a total of 70 dB, (assuming the frequencies given i.e. 1·8% off-tune). In a 4 MHz inter-block separation, the total would be about 80 dB because, although the isolator makes no contribution, the filter attenuation will have risen to about 50 dB.

Additional intermodulation suppression may also be obtained where isolators are used, by the insertion of a low pass filter between the isolator and the antenna, as

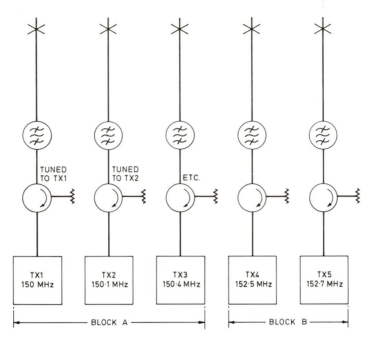

Fig. A10.8 Adding Coaxial Filters

shown in Fig. A10.9. Harmonics will be reduced by this method, enhancing the intermodulation suppression of the isolator configuration.

8. Types of isolator

Isolators are available from the manufacturers in either factory-tuned or tunable forms. Total tuning ranges do not appear to be usually specified in current manufacturers' literature. However, Microwave Associates do state that their tunable circulator/isolator H360, designed for the 77 to 88 MHz band, does in fact tune over the entire band.

9. Tuning procedure where tunable units are used

The tuning procedure specified by a reputable manufacturer is obviously superior to any ad hoc method and we therefore reprint the instructions given by Microwave Associates (Ref. 15).

The layout is shown in Fig. A10.10.

They state:

'It is important to note that all tuning should be done at a reduced power level. Do not tune the field tunable circulator at powers greater than 10 watts.

1.   Place the isolator in the transmission line in the forward direction (arrow pointing to antenna or load). Tune the input and output capacitors for minimum loss (maximum power) at the power meter connected as shown.

2.   Reverse the isolator in the transmission line so that the arrow is pointing towards the transmitter. Adjust the capacitor at the terminated port for minimum power (maximum isolation) at the power meter.

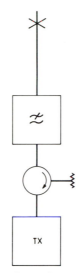

Fig. A10.9 Low Pass Filter in Antenna Feed

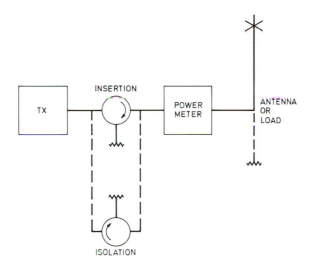

Fig. A10.10 Tuning Configuration

3.   Go back to step one and re-adjust if necessary.

The isolator is now tuned for that transmission line and that frequency.'

10.   Conclusions

Where a number of closely-spaced frequencies are used with inadequate antenna isolation, circulators or isolators are recommended as the primary means of reducing intermodulation.

Some care is needed in their use and, in special cases, where frequencies exceed their usable bandwidth, additional filtering may be necessary. This can take the form of bandpass, low pass or band reject filters.

Appendix 11. The Operation and Use of Duplexers

1. General

In duplex radio systems, the antenna system must permit simultaneous transmission and reception. The requirement can be satisfied in either of two ways:

(*a*) By using separate antennae for both the transmitter and receiver of the duplex pair.

(*b*) By combining the output of the transmitter and the input of the receiver in a device known as a duplexer which enables a single antenna to be used for both purposes.

This device is often referred to as a diplexer but generally the term duplexer refers to the unit intended for coupling a transmitter and a receiver into a single antenna; a diplexer is a similar device intended to couple two transmitters or two receivers into a single antenna.

The following notes are intended to illustrate the use of a duplexer, the principles of operation and the criteria for achieving the desired results.

2. Need for Isolation

2.1 Antenna Separation. The reason for using two antennae or a single antenna and duplexer, is to obtain isolation between the transmitter and receiver. In the design of any duplex system, it is most important to ensure adequate isolation to prevent the receiver from being adversely affected by its associated transmitter.

When using two antennae, one as an efficient radiator and the other an efficient pickup device, they must be suitably erected and spaced. The spacing must ensure that the receiver is not subject to transmitter signal voltages high enough to cause non-linearity or other undesirable effects. The required spacing varies with frequency separation between transmitter and receiver, transmitter power, etc.

2.2 Receiver Selectivity. The modern day communications receiver is presented with a VHF or UHF signal and systematically lowers the frequency, in one or two steps, as the signal passes through various stages of the receiver. As the frequency of the signal becomes lower, the passband of the receiver can be reduced (see Fig. A11.1). Finally, the received signal frequency is lowered to a point where the receiver circuits are able to pass an extremely narrow band of desired frequencies and reject adjacent frequencies by say, 70 dB. This is the overall selectivity of the receiver described on the equipment specification sheet. The overall selectivity of the receiver can be excellent but the front-end stages are relatively broad and cannot reject completely the strong signal from the transmitter, even though the transmitter and receiver frequencies may be separated by several MHz.

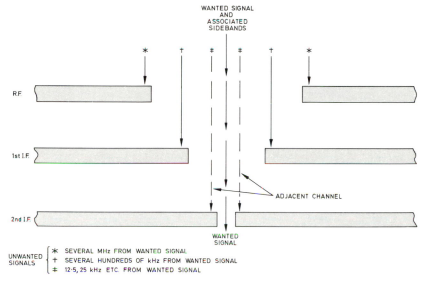

Fig. A11.1 The Receiver Selectivity Mechanism (assuming approximately equal signal levels)

2.3 Receiver Desensitisation. For optimum performance, critical RF voltage and current levels exist at certain points throughout the front-end stages of a receiver. If these levels are radically changed, the performance of the receiver will become degraded at its operating frequency and such effects as receiver desensitisation and spurious signal products arise (see Fig. A11.2).

2.4 Transmitter Noise. Transmitter noise also adversely affects the performance of a receiver. Ideally, the transmitter should confine all of its output power within a narrow band of frequencies on either side of the transmit channel. The bulk of the power is in fact confined within the assigned channel but some is radiated on other frequencies above and below the carrier frequency. The undesired radiation is referred to as transmitter noise. Filter circuits in the transmitter eliminate a considerable portion of the undesired radiation but enough noise energy is radiated to degrade the performance of a receiver operating several MHz away. The level of noise is greatest at frequencies close to the carrier frequency of the transmitter. (See Fig. A11.3 showing the transmitter noise spectrum and the effect of additional cavity filters. Without filtering, the receiver will be degraded. The effect is clearly less troublesome with wider frequency spacing.)

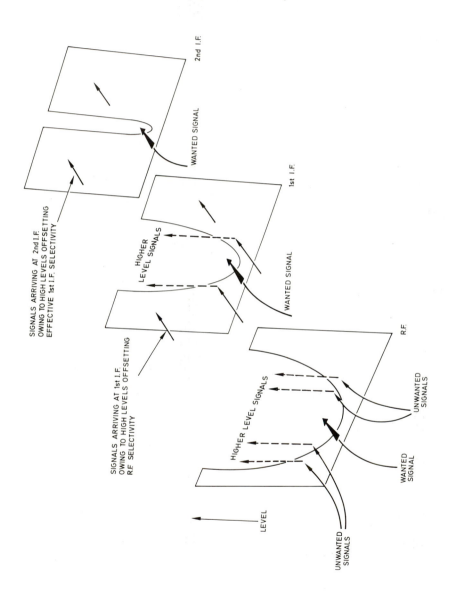

Fig. A11.2 Receiver Selectivity Mechanism with High-Level Signals

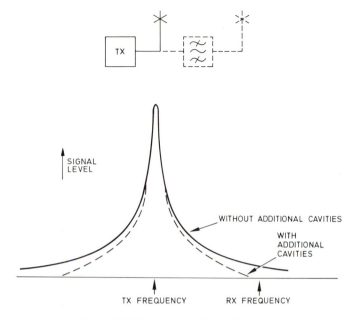

Fig. A11.3 Transmitter Noise Spectrum

Transmitter noise appears as "on-channel" noise interference to the receiver and cannot be filtered out at the receiver. It falls exactly on the operating frequency of the receiver and competes with the desired signal. Appendix 25 deals further with this subject.

Assuming that the frequency separation, power output etc., are acceptable, for correct operation, the remaining attenuation must be achieved by suitable antenna spacing to ensure that the level of transmitted signal at the receiver input is also acceptable. The smaller the frequency separation and higher the transmitter power the greater is the physical separation needed to achieve adequate isolation (minimum effective on-channel voltage from the transmitter). A point is ultimately reached when the cost of supporting antenna structures becomes a high proportion of total expenditure. Alternatively, the spacing available on one support becomes too great for the structure size. This is especially so when the support carries a large number of antennae.

These problems are particularly inhibiting when high gain antennae are involved and the various antennae affect one another increasingly.

At this point it becomes important to consider duplexers to minimise the effects.

3.   Advantages of a Duplexer

Generally, economy is seldom the reason for use of a duplexer. There are other, more important reasons:

3.1   Isolation. The proper duplexer will provide the necessary consistent isolation between the transmitter and receiver.

3.2   Antenna Pattern. Without a duplexer, a duplex system must have two antennae. Both antennae cannot be mounted at the same location on the support structure and the radiation patterns of the two antennae will probably be different. The coverage area of the transmitter could be somewhat different from the coverage area of the receiver. With a duplexer, the system uses a common antenna which provides the same pattern for both transmitter and receiver.

3.3   Tower Space. Good antenna sites are scarce and often crowded. It is always easier to find a place to mount one antenna than two.

4.   Disadvantages of a Duplexer

There are some disadvantages when using a duplexer. These should be weighed against the advantages when engineering a duplex mode system.

4.1   Cost. This aspect becomes important when high powered transmitters are considered. The high powers to be dissipated with quite normal losses (1 dB at the 1 kW level equals a loss of 200 W) require a large and expensive duplexer device. Higher power also necessitates greater rejection of the unwanted transmitter noise and this again complicates the assembly with a further increase in cost.

   At the lower power end of the scale, the cost of the simple duplexer is comparable with that of two antennae, particularly in the UHF part of the spectrum, and therefore generally acceptable in most applications of this nature.

4.2   Unnecessary or Uninformed Adjustments. Possibly one of the hazards of a duplexer is its adjustment by persons not familiar with the technique. All circuits should be sealed against such adjustments and, if possible, attention should be drawn by a label warning all against such action.

   This is particularly true of a band reject duplexer which can only be adjusted in the field with the aid of additional test equipment. The bandpass duplexer can be adjusted using the equipment to which it is connected, but even this requires expert knowledge and should not be attempted by the uninitiated.

4.3   Ageing. Some ageing or settling-down time is usually necessary after the initial installation. This is caused by mechanical movements of the various parts of the device and it is recommended that some months after installation the duplexer tuning be checked. The band reject duplexer (particularly the unit operating at close frequency spacing) usually shows the effect of drift more than the bandpass unit and consequently may require more than one readjustment in the early days of use.

4.4   Loss. Loss means heat and introduces several problems with duplexers.

   As the power loss increases, the duplexer size must be increased to enable the heat to be dissipated with minimal ill effects. The ageing effects and the associated detuning, excessive heat and lost power can sometimes cause breakdown of the sections.

   The disadvantages given in section 4.1 to 4.4 all indicate that the merits of the duplexer are greatest with lower power transmitters but, at high power, it can be both expensive and complicated if it is to operate with minimal ill effects. Nevertheless, duplexers are available for high power transmitters and are used with success. This is particularly true where expensive high gain antennae are included in the system and where considerations of mounting and size encourage reduced antenna complexity.

5. Duplexer Requirements

There are two distinct types of duplexers used in two-way radio communications: the bandpass duplexer, and the band-reject duplexer. Each type has its advantages and disadvantages and both types are discussed in the following paragraphs. The selected duplexer must provide certain functions for optimum performance.

A duplexer must:
- (i) be designed for operation in the frequency band in which the duplex system operates.
- (ii) be capable of handling the output power of the transmitter.
- (iii) be designed for operation at, or less than, the frequency separation between the transmit and receive frequencies.
- (iv) provide adequate rejection to transmitter noise occurring at the receive frequency.
- (v) provide sufficient isolation to prevent receiver desensitisation and production of spurious signals.
- (vi) offer as little loss as possible to the desired transmit and receive signals.

6. Losses through the Duplexer

The output signal from the transmitter and the incoming signal to the receiver are both reduced by losses in the duplexer. These losses are usually referred to as "insertion loss, transmitter to antenna" and "insertion loss, receiver to antenna" on the duplexer specification sheet. Generally, the insertion loss will increase as the separation between transmit and receive frequencies is decreased. The losses could be between 0·5 and 2 dB in both the transmit and receive legs.

7. The Bandpass Cavity

Before describing the theory of the bandpass duplexer, the characteristics of the bandpass cavity are best considered. A bandpass cavity is a device which serves as a filter of radio frequencies. It has the ability to let a narrow band of frequencies pass through while frequencies outside of the narrow band are attenuated. Energy is normally fed into the cavity by means of a coupling loop, which excites the resonant circuit formed by the inner and outer conductors. A second loop couples energy from the resonant circuit to the output. The coupling coefficient of the loops and the "Q" of the cavity determines the selectivity and the narrow band of frequencies that pass through with only slight loss are those reasonably close to the resonant frequency of the cavity.

The selectivity of a typical bandpass cavity is illustrated by the use of the frequency response curve shown in Fig. A11.4. The curve indicates the amount of attenuation the cavity provides at discrete frequencies above and below the normalised frequency. It also indicates the amount of insertion loss caused to the desired (pass) signal at the resonant frequency of the cavity.

If a single cavity will not provide enough rejection to an undesired signal, additional cavities may be connected in series to improve selectivity. Additional cavities slightly increase the insertion loss at the desired frequency but give substantial improvement in overall selectivity.

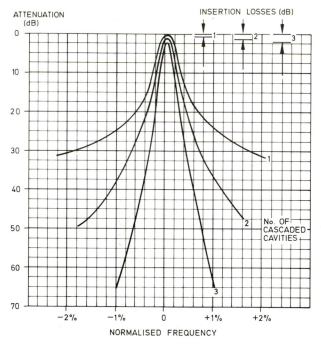

Fig. A11.4 Typical Responses of Cascaded Cavities

8.   The Bandpass Duplexer

This is so called because it is made up of two or more bandpass cavity filters interconnected in a duplexer configuration. One or more cavities are placed in the
transmitter section of the duplexer and tuned to pass the transmit carrier frequency
and associated sidebands. In a similar manner, the bandpass cavities in the receiver
section of the duplexer are tuned to pass the narrow band of frequencies at the
receiving frequency.

   The output of the transmitter is fed through the bandpass cavities in the transmitter section of the duplexer to the antenna. Since the cavities are resonant to the
transmit frequency, they allow the transmitter carrier and sidebands to pass
through with very little loss. But the energy on all other frequencies is attenuated.
Transmitter broad band noise energy at the receiver frequency that would normally
be radiated from the transmitter is reduced in the cavities which also reduce
transmitter noise on other frequencies. Other receivers in the area can also benefit
from the noise reduction feature of the bandpass duplexer. Fig. A11.3 illustrates
how the transmitter noise is reduced at frequencies removed from the transmitter
channel. The received signal from the antenna is fed through the bandpass cavities
(usually two or more) in the receiver section of the duplexer, to the receiver. These
cavities are resonant at the receive frequency and the desired signal passes through
the cavities with only slight loss. All other frequencies on either side of the resonant
frequency of the cavities are attenuated (Fig. A11.5). Essentially, the front-end
selectivity of the receiver has been improved. These bandpass cavities also help in
protecting the receiver from other nearby transmitters.

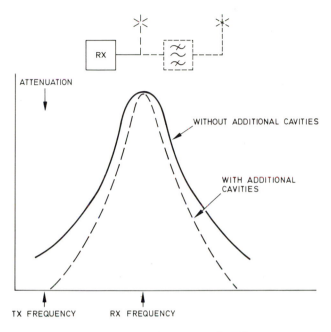

ATTENUATION

WITHOUT ADDITIONAL CAVITIES

WITH ADDITIONAL CAVITIES

TX FREQUENCY   RX FREQUENCY

Fig. A11.5 Typical Receiver Front-End Response

A cable harness is used on a bandpass duplexer to interconnect the numerous cavities in the two sections of the duplexer. The cables are cut to special length to make the harness a matching device.

The bandpass duplexer is relatively simple to install and requires practically no maintenance. If desired, additional cavities can be included in either section to obtain additional isolation. It is not suitable for use in duplex systems with close spacing between transmit and receive frequencies. The bandpass duplexer curve (Fig. A11.6) shows that, at reasonable insertion loss levels, it cannot effectively attenuate frequencies near the resonant frequency. Maximum attenuation occurs only at frequencies far removed from the resonant frequency, and limits the use of the bandpass duplexer to systems with wide frequency spacing.

The degree of receiver selectivity improvement required by the use of cavities must be determined by the frequency separation of the transmitter and receiver and must be sufficient to protect the receiver against desensitisation.

9. The Band-reject Filter

The band reject filter attenuates a band of frequencies while allowing all other frequencies to pass through with only slight loss. Energy at the resonant frequency (the reject frequency) is absorbed.

Maximum attenuation occurs at the resonant frequency of the filter as shown in Fig. A11.7. Unlike the bandpass cavity, the band-reject filter provides a given amount of attenuation at resonance regardless of the separation between the pass and reject frequencies. The filter can be tuned to place the narrow band of rejected frequencies several MHz from the desired pass frequency, or reasonably close. Minimum frequency separation is limited only by the amount of loss that can be

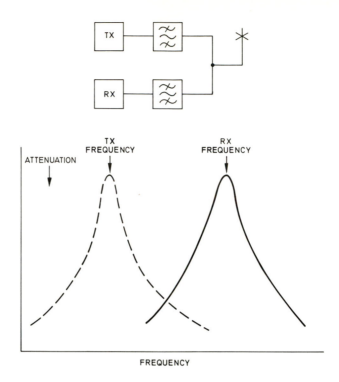

Fig. A11.6 Response of Typical Band-pass Duplexer

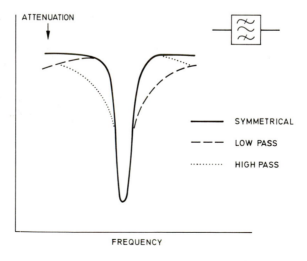

Fig. A11.7 Response of Typical Band-Reject Filter

tolerated at the desired frequency. With the use of stubs, the same filter can be made to provide one of several different frequency response curves (Fig. A11.7). Filters can be added in series to obtain additional attenuation to an undesired frequency, with two filters providing about twice the attenuation to the undesired frequency as a single filter. The most important feature of the band reject filter is the steepness of the frequency response curve. This permits the filter to provide maximum attenuation to an undesired frequency which is quite close to the desired frequency.

10.   The Band-reject Duplexer

Band-reject duplexers consist of band-reject filters (notch filters) interconnected in a duplexer configuration. One or more filters are placed in the transmitter section of the duplexer and are tuned to reject a band of frequencies around the receive frequency.

Similarly the filters in the receiver section of the duplexer are tuned to reject a band of frequencies around the transmit frequency (see Fig. A11.8). This is exactly the opposite function to a bandpass duplexer.

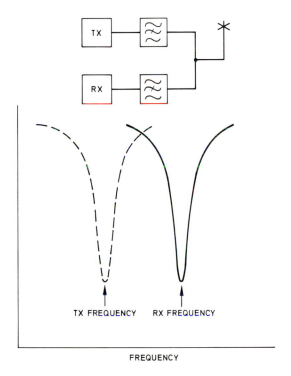

Fig. A11.8 Response of Typical Band-Reject Duplexer

The transmitter output is fed through the band-reject filters in the transmitter section of the duplexer to the antenna. Since these filters are tuned to the receive frequency, they accept and absorb the transmitter noise energy being radiated by the transmitter that would normally appear at the receive frequency. The energy at all other frequencies transmitted is passed by the filters with less attenuation (see Fig. A11.9).

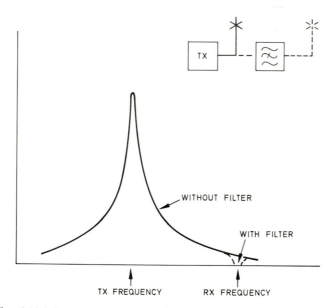

Fig. A11.9 Reduction of Transmitter Noise by Band-Reject Filter

The band-reject duplexer takes a more direct approach in protecting the receiver from noise radiated by its associated transmitter. In contrast to the bandpass duplexer, the band-reject type does not alter the overall transmitter noise output. Instead it selectively rejects transmitter noise at the critical band of frequencies near the receive frequency.

The incoming signal from the antenna is fed through the band-reject filters in the receive section of the duplexer to the receiver. These filters are tuned to the transmit frequency and absorb the transmitter energy, at and near the transmit frequency, that might normally be fed to the receiver. The desired incoming signal and energy on all other frequencies are passed by the filters with only slight attenuation. Thus the undesired energy is rejected and reduced to a level where it can no longer cause receiver desensitisation or spurious signal production. Unlike the bandpass duplexer, the band-reject type does not change the overall front end selectivity of the receiver. Instead, it changes only a portion of the selectivity and makes the receiver unresponsive to the critical band of frequencies at and near the transmit frequency (see Fig. A11.10).

The cable harness in a band-reject duplexer interconnects the filters in the two sections and ensures correct matching in the two branches and the equipment.

The band-reject duplexer is used more than other types owing to its compact size, low insertion loss and excellent isolation features. It is often used at wide frequency

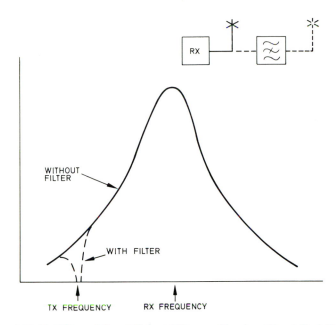

Fig. A11.10 Effect of Band-Reject Filter on Receiver Front-End Response

spacing and almost exclusively at closer frequency spacing. Many of the band-reject models include shorter helical cavity filters which can usually be mounted inside the radio equipment cabinet.

11.   Other Duplexer Types

Other types of duplexers exist, but they all operate on the principle of the bandpass duplexer or the band-reject duplexer or a combination of the two. Some use coils and capacitors in an electronic circuit. These are quite small and are generally limited to use in mobile units. Others include bandpass cavities in one section of the duplexer and band-reject filters in the other, or a combination of both in each section, in order to achieve a specific isolation characteristic.

12.   Use of Duplexers as Combiners

Duplexers can also be used to couple two transmitters, two receivers, or two single frequency simplex stations into a common antenna. The duplexer is then usually called a diplexer or combiner.

13.   Power Rating

A duplexer is rated at a particular power level. The power rating shown on the duplexer specification sheet normally includes some safety margin but the specified power level should not be exceeded if normal performance is to be achieved. Excessive power can also cause excessive temperatures, detuning, voltage breakdown and physical damage to the duplexer.

14.   Temperature Effects

Duplexers are expected to remain tuned and provide a specified performance over an extremely wide temperature range, often –30°C to +60°C. This presents a problem since conventional metals will contract and expand when exposed to temperature changes. To solve this problem, most duplexers are temperature compensated to ensure that the resonant frequency of the cavity filters remains stable despite a change in temperature. Several methods of temperature compensation may be employed, the most common being the use of INVAR metal at critical points within the filters. The temperature coefficient of INVAR is practically zero and the metal is virtually unaffected by changes in temperature. The specifications of a duplexer usually cover the performance characteristics of the unit operating at the temperature extremes.

15. Frequency Separation

Duplexers are usually rated as being suitable for use at a certain minimum frequency separation such as "3 MHz or more". If operated at closer frequency spacings than recommended, the duplexer will probably have inadequate isolation, excessive insertion loss at the desired frequencies, or both. Duplexers must be suitable for the specific frequency separation when used with a particular duplex station.

Appendix 12. Single Frequency Simplex, Two Frequency Simplex and Duplex

1. General

This Appendix contains notes on the three categories of operation generally employed in the UHF and VHF bands:

(a) Single frequency simplex. In this mode the transmit and receive frequencies are identical and consequently, with the exception of certain techniques described later, the receiver is not operative during transmission periods. The operation is basically a "press-to-talk" method.

(b) Two frequency simplex. Here the simplex method employed utilises two frequencies separated by an amount sufficient to permit satisfactory system operation when subjected to certain site restrictions described later. Again the operation is basically a "press-to-talk" method.

(c) Duplex. In this method both the transmitter and receiver can operate simultaneously, permitting a normal two-way conversation to take place. No "press-to-talk" is needed.

Each category has advantages and disadvantages and the method finally adopted for any particular system must be based upon the full operational requirements of that system, such as frequency economy, siting, type of user, methods of usage, equipment types and frequency switching bandwidth.

The following paragraphs discuss the problems in greater detail.

2. Frequency Economy

Fig. A12.1 shows frequency allocations at a communal site for (i) single frequency simplex and (ii) two frequency simplex and duplex. The irregular channel spacing is for the avoidance of on-channel third order intermodulation products $(2a - b)$.

2.1 Single Frequency Simplex. In a simple system using an isolated transmitter and receiver, this method is generally satisfactory as it permits fixed-to-fixed, fixed-

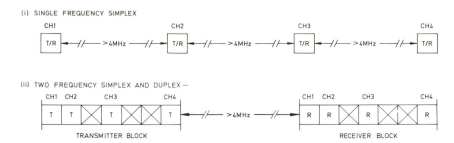

Fig. A12.1 Frequency Economy Comparison

to-mobile and mobile-to-mobile conversations, with everyone within range able to talk and listen to all other units. Under these circumstances one frequency only is required and would appear at first sight to be the most economical frequency utilisation method.

Undoubtedly this mode operates with complete satisfaction in large open tracts of country, large areas of water, air communication etc., but if there is a requirement for adjacent siting of more than one channel, the apparent saving in frequencies is lost.

From Appendix 6, it can be seen that when receivers are located on the same site as transmitters, it is essential that a gap of over 4 MHz exists between the frequencies of any receiver and any transmitter operating at the same time. Obviously with a single transmitter/receiver this is not necessary when operating simplex, but immediately two systems are located at the same site, the frequencies of the two systems must be so separated to permit both systems to operate satisfactorily and independently.

Increasing the co-sited units to three will necessitate a further channel, spaced from the others by a similar amount. However, a further problem now arises. This is on-channel intermodulation requiring the channels to be irregularly spaced to prevent interference.

It can therefore be seen that, far from economising in frequencies, the use of single frequency simplex merely wastes the frequency spectrum by requiring excessive channel spacing.

2.2   Two Frequency Simplex. Examining the general frequency requirements for two frequency simplex, it can be seen that the initial simple system does in fact appear less economical in its use of the frequency spectrum when compared with single frequency simplex, as it uses two frequencies spaced by over 4 MHz instead of one frequency as with the single frequency simplex mode.

However, by resorting to the use of two blocks of frequencies, separated, for example, by 4 MHz, an entirely different picture emerges in an area where a concentration of radio channels is required. By locating the frequencies of all fixed transmitters in one block and similarly the frequencies of the fixed receivers in the second block, it can be seen immediately that, with certain reservations concerning intermodulation generation, the number of equipments located in a small area will only be limited by the number of adjacent channels available in the blocks. Obviously, therefore, co-siting of equipment merely becomes a routine task of ensuring that normal engineering principles are observed to minimise interference.

2.3   Duplex Operation. This mode of operation is, in most respects, similar to two frequency simplex. The use of two blocks in a basic system enables the number of channels available at a communal site to approach the number of adjacent channels available in any block. Only limitations such as those imposed by the generation of on-channel intermodulation products will restrict the channel allocations.

When duplex operation is compared with two frequency simplex, third order intermodulation based on the sequence a + b – c may be seen to be more troublesome. (Appendix 6 Annexe B). Various methods of reducing this can be employed and mainly consist of ensuring that the sites housing the base stations are located outside the area of the main mobile concentration. This reduces the possibility of high signal levels from the mobiles causing non-linearity in the base station receivers.

Nevertheless, even with this additional shortcoming, channel economy can be considered similar to that achieved with two frequency simplex.

3.   One Method of Minimising Frequency Spectrum Usage in Single Frequency Simplex Systems

As can be seen in Section 2 above, systems located on one site can only operate satisfactorily when a suitable barrier exists between any transmitter and any receiver on that site. Fig. A12.1 shows that if the "barrier" is in the form of a frequency separation as with two frequency simplex, then isolation and satisfactory operation can be achieved on a given site. The figure also shows that two frequency simplex is the only method possible on a single site if frequency spectrum economy is to be achieved in a practical manner.

However, an alternative method to assist frequency economy is possible and is of particular use in systems where single frequency simplex is essential, such as air-to-ground communications and marine operations.

In all communication systems, in order to achieve satisfactory operation, the receiver must never be subjected to an input voltage which is high enough to produce non-linearity in the early RF stages. It has been shown that it is possible to avoid this condition by adequate frequency spacing, as used in two frequency simplex systems. It can also be obtained by physical separation between transmitters and receivers when the single frequency simplex mode is employed.

Thus, if an attenuation between the transmitter output and the receiver input of say, 80 dB is required to prevent non-linearity, this figure can be achieved by front-end receiver selectivity using two frequency simplex with a spacing of, for example, 4 MHz, or by physical spacing when the channels are closely spaced. A combination of both is often used for special requirements described later.

Obviously, physical spacing only can be employed in systems such as air-to-ground communications and marine applications where a number of channels are in use with single frequency simplex. Distances between separated sites in excess of 1 km are usual.

Fig. A12.2 shows the method in block form.

Unfortunately, whilst being highly satisfactory for large systems such as are employed for airport communications, the separation of transmitter and receiver

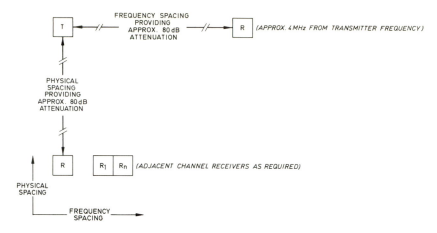

Fig. A12.2 Comparison between Frequency Spacing and Physical Spacing

sites is obviously more costly than a single site and, additionally, necessitates land-lines or radio links for operation from one control point. It is quite often usual to co-site the control complex with either the receiver or transmitter equipment but, even so, the additional site is expensive and its cost must be weighed against the operational need for single frequency operation.

4.   The Use of Separated Sites With Two Frequency Simplex and Duplex Systems

Whilst, at first sight, the use of separated sites with two frequency systems would appear to offer no advantages, it can be seen after examination as a method to simplify mobile unit complexity.

If one considers the case where a particular system requires, as an operational necessity, not only base-to-mobile communication but also mobile-to-mobile communication when out of range of the base station, then three solutions are possible:

(*a*)   By the use of a special mobile unit having broad RF bandwidth characteristics enabling either or both the transmitter and receiver to switch several MHz from a two frequency mode of operation to single frequency operation.

This type of mobile is available but, being of a special design, is naturally more expensive than the standard narrow bandwidth unit.

(*b*)   By the use of a special mobile unit having an additional receiver RF head to enable two frequency simplex operation between the transmitter and the first receiver combination, with the second receiver combination permitting single frequency operation by being tuned within frequency switching distance of the transmitter.

This again is a more expensive solution than a standard type of mobile unit.

(*c*)   A standard mobile unit having a maximum switching bandwidth of $\pm 0\cdot2\%$ of the nominal frequency in the case of VHF and $\pm 0\cdot1\%$ at UHF is employed.

By separating the base station elements and using two sites, the requirements outlined in 3 above are achieved. It is then possible to reduce the frequency spacing between the transmitter and the receiver to a point where by switching the frequency of either or both transmitter and receiver in the mobile by not more than 0·2%, the mobile can be made to operate in the single frequency mode. Fig. A12.3 shows the detail. The scheme can also be implemented by using high-Q filters at the fixed station, if the two frequency spacing is adequate.

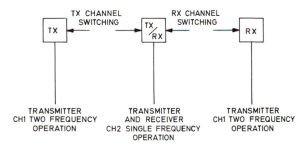

Fig. A12.3 Operation from Separated Base Station Sites

Obviously, with the last method (*c*), the cost of providing separated sites is an appreciable part of the total system cost, but in a system employing large numbers of mobile units the effective increase in outlay becomes less as the quantity of mobiles increases. In the other two cases (*a*) and (*b*), the extra outlay will be directly proportional to the quantity of mobiles in use.

If one analyses the three methods by showing total outlay against quantities of mobile units, it will be seen that a cross-over will occur, above which the last case (*c*) shows a definite economic advantage.

For economy in site costs, it should be appreciated that the distance between sites in this particular case obviously does not require to be as great as with single frequency simplex unless definite on-channel inter-modulation products are involved. The site separation must, of course, be sufficient to eliminate possible blocking effects in the receiver, but as the frequency separation between the transmitter and the receiver in the two frequency configuration is based on the mobile switching bandwidth, it can be seen that, at 150 MHz, the transmitter and receiver frequency switching can certainly extend over 600 kHz, and with offset tuning, up to 1·2 MHz. This is adequate to provide sufficient isolation and to allow the two sites to be within the confines of a reasonably large building area, thus eliminating rented lines, radio links etc.

5. Break-in Operation

In any system, small or large, shared or exclusive, the essence of good communication is brevity. In large or shared systems, every message must be short as possible in order to make the maximum use of the time available.

Nevertheless, even with short messages it is often necessary for a mobile to break in with an urgent message and, therefore, the best system is necessarily the one permitting such break-in without causing system confusion.

Examining the various modes of operation it can be seen that the following conditions exist:

5.1 Single Frequency Simplex. In this mode the channel occupancy approaches 100% during an exchange of messages and therefore very little opportunity exists for another station to break in during such an exchange, unless it is an appreciably stronger signal. This, of course is operationally unsound.

5.2 Two Frequency Simplex. Although, in this case, the mobile transmitting channel is unoccupied when a message is being transmitted by the base station, the fact that the base station is not receiving, places the system in the same general category as single frequency simplex from the point of view of break-in.

5.3 Duplex. With full utilisation of both transmit and receive channels during a conversation, break-in is impossible unless the station breaking in is very close to the base station (again operationally unsound).

Fortunately, by combining 5.2 and 5.3, an ideal system can be operated. With this method, the mobile unit operates in a conventional simplex manner, whilst the base station operates in a duplex mode.

With this method it can be seen that, whilst transmitting, the base station receiver is listening out and, in any emergency, a mobile or other outstation can call and be heard by the base station operator. Of course, this method can lead to operational misuse if calling is allowed to take place without restriction but, nevertheless, it does provide the opportunity to do so without the jamming effect which occurs with the methods outlined previously.

6. Telephone Connection

Connection to either an internal or external telephone system must necessarily be achieved by one of the following methods:

6.1 When the base station operates in the simplex mode, whether single or two frequency, it is of course necessary for the "press-to-talk" facility to be achieved in some manner or other.

Three alternatives exist

(a) Operator intervention. Here the operator extends the radio system – on a two wire basis – into the telephone system concerned and operates the send/receive switching locally, whilst monitoring the conversation. Obviously, standard radio techniques indicating the end of the message, etc. must be followed for the operator to be able to achieve a satisfactory procedure.

(b) By extending the "press-to-talk" facility to the telephones involved in a given building. In certain cases this method is quite satisfactory and, when only a small number of telephone extensions in a limited area are involved, the method is not expensive.

   However, its expansion is limited and obviously distance is the main deciding factor.

(c) *VOX (Voice Operated Switching)*

   This method is used quite often and is the obvious choice when extending simplex radio circuits to telephone instruments, whether in the same building or externally.

6.2 When the fixed station operates in the duplex mode, the problems associated with extending "press-to-talk" switching disappear. Over short distances no problems really exist. Over the longer distances, however, the question of levels, hybrid balancing, etc. raise difficulties in its use but, even so, it is obvious that the use of duplex base stations is the only method by which successful telephone radio operation can be assured.

Apart from a modification to the operational procedure, the system applies equally well whether simplex or duplex mobiles or outstations are used.

7. Talkthrough Repeater Operation

One of the most widely used of the facilities available is only made possible by including two frequency (duplex) operation. This facility is known as talkthrough.

It operates by connecting the receiver audio output into the transmitter modulator input at a carefully adjusted level. Then, by utilising a "make" contact on the squelch relay and connecting this contact to trigger the transmitter, the automatic retransmission of signals is accomplished.

Since both the receiver and transmitter will be operating when a message is being retransmitted, this method can be considered as operation in the duplex mode. Thus, it is essential to fulfil the requirements outlined in the previous paragraphs.

Two types of talkthrough station configuration can be considered:

(*a*) The conventional talkthrough repeater using a transmitter-to-receiver spacing in excess of 4 MHz. Here the transmitter and receiver are co-sited with or without other units and, apart from normal engineering and frequency allocation problems, no difficulty should arise.

(*b*) The specialised closely spaced talkthrough repeater with a transmitter-to-receiver spacing close enough to permit both two frequency and single frequency simplex in outstations using the system.

Here the separated site method outlined in 4 above can be adopted and is essential if a number of cosited talkthrough stations using close frequency spacing are involved. However, when only one talkthrough station in a particular band is needed at a given site, an alternative system using a high Q duplexer can be used. With such a filter it is possible to achieve a transmitter-to-receiver spacing approaching 500 kHz at a nominal carrier frequency of 150 MHz. This, it can be seen, is well within the $\pm0.2\%$ switching bandwidth of mobile equipments, permitting switching functions (for fixed to mobile, mobile to mobile operation) similar to those described in 4 above.

8. Side Tone Monitoring

This is mainly used in ground-to-air single frequency simplex systems where separated sites for transmitters and receivers are necessary.

Obviously, if conditions exist where, by the use of two sites, the signal level from a transmitter into its companion receiver on the same channel is considerably below the receiver blocking level, it will be unnecessary to switch off the receiver during transmission. In this state, therefore, any outgoing transmission, as well as incoming signals, will automatically be received by the equipment concerned.

Three advantages result from this method:

(*a*)   Side tone will automatically be available at the output of the receiver.

(*b*)   Tape recordings can be made of both outgoing and incoming conversations.

(*c*)   A continuous check can be made of any outgoing transmission and therefore any failure of a transmitter will be noted immediately.

To achieve side tone monitoring with a two frequency simplex system, an additional frequency changing device is necessary to convert the transmitter frequency to that of the receiver and to attenuate the resulting output to a level suitable for operating that unit.

However, this method is mainly used for test purposes in specialised two frequency systems and is seldom fitted permanently.

Appendix 13. Talkthrough Operation – Advantages and Limitations of Reverse Frequency Repeaters

1. General

The use of talkthrough as a simple, relatively cheap and effective repeater has been known for many years. It is restricted usually to systems using the two frequency method of frequency planning, although special systems containing a mixture of single and double frequency allocations often employ a sophisticated talkthrough repeater in spite of filtering problems introduced by using closely-spaced channels.

Although simple, talkthrough has limitations and this Appendix outlines not only the techniques but also the precautions in its use and the circumstances where it is inadvisable.

2. Methods of Applying and Controlling Talkthrough

Talkthrough can be applied and controlled in numerous ways, many of which are determined by operational characteristics. A number of methods are influenced by the geographical and physical layout of the area, whilst several are dictated by telephone line limitations etc. The following are some of the more likely alternatives:

2.1 Unattended remote free-running talkthrough station. This is the commonest method and usually consists of a station on a high remote site. The system is triggered on receipt of a carrier. Fig. A13.1 shows the arrangement. The two antennae may, of course, be replaced by one with a duplexer. The audio output is connected to the modulator via a suitable attenuator.

2.2 Unattended remote station controlled over a telephone pair. Several methods can be used with this technique:

(*a*) DC controlled, with application of DC between one leg and earth applying the talkthrough control signal (see Fig. A13.2 (Type A)).

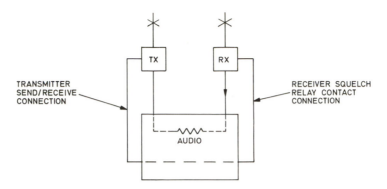

Fig. A13.1 Unattended Remote Free-Running Talkthrough Station

(*b*)   DC controlled, with application of DC between one leg and earth inhibiting talkthrough. This method has the advantage of line fail characteristics, where, with the disconnection or failure of the telephone pair, the system reverts to talkthrough and permits the overall system to continue operating using mobile units until the line is restored (see Fig. A13.2 (Type B)).

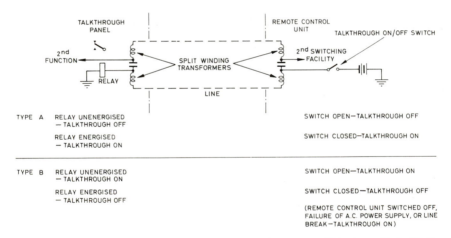

Fig. A13.2 Control of Unattended Remote Station over a Telephone Pair (D.C. Method)

(*c*)   Where DC cannot be used due to P & T regulations, one or more tones can often be used to switch the remote transmitter and receiver to talkthrough. The tone can be filtered above a restricted audio band or an in-band single or multiple tone burst can be used. Fig. A13.3 shows the general arrangement. Note that, by reversing the relay contacts, in A13.3 (i), talkthrough can occur on either the application or removal of the tone.

2.3   Unattended remote station controlled over a radio link. As with the previous method, several alternatives are possible with this type of system:

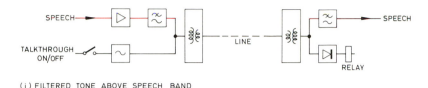

(i) FILTERED TONE ABOVE SPEECH BAND

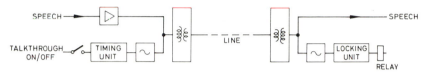

(ii) IN-BAND TONE BURST

Fig. A13.3 Control of Unattended Remote Station over a Telephone Pair (Tone Method)

(a)   *Link fail method*

This is possibly the most useful of the alternatives. It relies on the establishment of a sequence of events within a predetermined period of time. Provided this sequence occurs, the system is prevented from switching to talkthrough. If, however, there is a break in the chain, owing to either a fault or a deliberate switching function, the system changes to talkthrough operation, permitting an emergency or an alternative method of control.

This system has been designed using either DC or tone triggering. Fig. A13.4 shows the general layout of this arrangement. Tones would normally be allocated a section of the audio spectrum above the speech band. The sequence of operation is as follows:

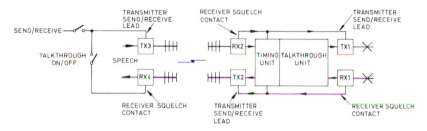

Fig. A13.4 Control of Unattended Remote Station over a Radio Link (Basic Method of Operation of Link Talkthrough – D.C.)

A signal at Rx 1 triggers Tx 2, Rx 4. If the talkthrough switch is closed then Tx 3 and Rx 2 are triggered. If this occurs within a preset time, the timing unit prevents talkthrough. If the talkthrough switch is open, or if the link is faulty, Rx 2 is not triggered; therefore the system operates in the talkthrough mode with Rx 1 triggering Tx 1 directly and an audio connection made between Rx 1 and Tx 1. A delay of approximately 5 seconds is included to prevent the mode changing back after each transmission. Normal operation from control is possible by closing the talkthrough switch and operating the send/receive switch in the normal manner.

(*b*)   *Control over-ride method*
In this method, the actual talkthrough switching is automatically achieved at the remote site by the normal send/receive function. Whilst in the receiving state, the system also operates on talkthrough. When control operates the send/receive function, talkthrough is removed, thus permitting normal operation from control without the possibility of a mobile message being superimposed on the speech from control. Talkthrough is re-instated when the transmit function is removed. Fig. A13.5 shows the arrangement. With this system, control can always break into the system even when outstations are speaking.

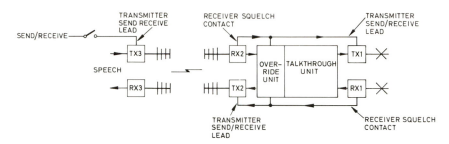

Fig. A13.5 Control of Unattended Remote Station over a Radio Link (Basic Method of Operation using Control Over-Ride – D.C.)

The operating sequence is as follows:

A signal at Rx 1 normally is fed to Tx 2 and Tx 1. Tx 1 is triggered by Rx 1 and the system operates on talkthrough with monitoring at control. If control transmits, the over-ride unit removes the talkthrough condition during the period of transmission. Talkthrough includes a 5 seconds delay to prevent the system collapsing after each talkthrough operation.
In both this method and (*a*) above, the tone alternative uses out of band tones for send/receive instead of squelch operation and the link usually operates continuously.

(*c*)   *Tone control*
As with control over a telephone pair, control over a radio link can also be achieved by a tone method. Again, this can be a continuous tone or tone burst using a single or multiple tone sequence. The continuous tone(s) would normally be in a part of the audio band above a restricted speech segment.

2.4   Unattended tone-operated station. Where the conventional free-running talkthrough station is prohibited by local regulations or where, for operational reasons, a free-running system is not required, there are several possible methods by which the repeater can be put into operation when desired:

(*a*)   *Sub-audio tone triggering*
This method operates from control or a mobile by accompanying the standard speech transmission with a continuous low level sub-audio tone below 300 Hz.

The reception of this tone switches the repeater to talkthrough and holds it in this condition throughout the transmission. The answering station, which also transmits speech plus tone, re-instates the condition, as does each subsequent transmission from mobiles of the same system.

Any transmission from other users lacking the tone, will not trigger the talkthrough repeater. Fig. A13.6 shows the general method. Again a single antenna may be used with a duplexer.

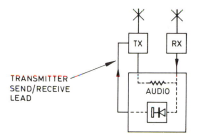

Fig. A13.6 Unattended Tone-Operated Station (Sub-Audio Tone Triggering)

(*b*) *Tone burst triggering*

With this method a burst of in-band tone(s) switches the repeater to talk-through. This condition is held,

(i)   as long as a carrier is received, and

(ii)   provided any individual break in transmission at the changeover between two users does not exceed a predetermined time (say 5 seconds).

In this system, each new transmission re-instates the system and re-times the break delay. Fig. A13.7 shows this. The holding unit maintains the talk-through condition after priming by the tone sequence. The period when talkthrough is maintained is adjustable and is reset by a carrier being received within that period. After collapsing due to the predetermined period of "no-carrier" being exceeded, the talkthrough condition can only be restored by a new tone burst sequence. Provided each transmission is within the predetermined time, carrier only will maintain the condition.

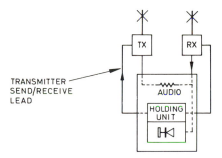

Fig. A13.7 Unattended Tone-Operated Station (Tone Burst Triggering)

The advantage of this system, where the initiation of a message sequence always originates from certain stations, is that tone burst encoders need to be fitted to the originating stations only.

3. Advantages of Talkthrough

(*a*)   Undoubtedly the main advantage of talkthrough is its simplicity. Talkthrough can be added to any base station using the two frequency method of operation, enabling mobile-to-mobile operation to be optionally achieved.

(*b*)   The remotely sited free-running talkthrough station is ideal as a repeater in areas with no electricity supply. Using solid-state equipment, power consumption can be minimised, the main drain being that due to the receiver with occasional periods of repeater operation. With the more conventional repeater using separate links, the current drain is obviously much higher.

(*c*)   No additional frequencies are needed over and above those required for the basic system.

(*d*)   Antenna requirements are minimal, only two being required normally, or alternatively, a single antenna if a duplexer is fitted.

(*e*)   Line fail talkthrough permits two alternative methods of operation. The advantage of emergency standby being available is obvious.

(*f*)   Due to the general simplicity of the system and the possible reduction in antennae the talkthrough repeater lends itself to installation in vehicles.

(*g*)   Physical space occupancy is minimal compared with a conventional link type repeater.

4. Disadvantages of Talkthrough

(*a*)   In the free-running mode, the problem of widespread interference due to the triggering by random distant outstations and mobile equipments on high sites can often prove to be a major stumbling block.

(*b*)   In an endeavour to overcome the interference problem, tone triggering is often considered. However, in spite of its effectiveness, it must be added to each outstation or mobile if full access to the system is intended.

(*c*)   Alternatively (and this method is generally preferred and is often mandatory) control over talkthrough must be invested in the main control operator. This increases the responsibility of the operator and introduces an additional function at control.

(*d*)   The use of more than one free-running talkthrough station on the same frequency in the same area is not normally possible. This is due to the ability of the control station to trigger two or more repeaters simultaneously, thus subjecting mobiles over a wide area with the product of two or more signals, the result of which can range from capture to complete unintelligibility.

Similarly control, when subjected to signals from two or more repeaters will suffer the same effect either capturing one or being subjected to severe distortion. If a mobile is received by two or more repeaters, therefore, it does not follow that the best signal will arrive at control; indeed the signal arriving at control may not be at all intelligible although the mobile signal into one or even both of the repeaters may be perfectly satisfactory.

(*e*) To overcome such problems as in (*d*) above, the use of separate channels is often suggested. Unfortunately this introduces problems of frequency availability and, additionally, increases the cost of providing multi-channel mobiles together with the operational hazard of area channel changing.

(*f*) Alternatively, the same channel can be used throughout the system but with different tone combinations to trigger the individual repeaters. This complicates the mobile units with a resulting increase in overall system cost. This system still requires area switching with the attendant hazard of mobile operators forgetting to make the change.

(*g*) The duty cycle of a talkthrough station is usually quite high. In fact, in a heavily loaded system, the transmitter can be radiating for a period approaching 100% of the time. Any talkthrough system must be designed bearing this in mind and the rating of the transmitter considered in this light.

(*h*) If several stations operate in the talkthrough mode on a single site, this can produce a high level of interference, due to the high duty cycle of each talk-through repeater. Intermodulation based on a + b – c is a problem under such conditions and this form of interference often limits the talkthrough mode to a very small number of repeaters per site, unless special care is taken when allocating frequencies.

To reduce the effect, it is often advantageous to separate the transmitter and receiver sites so that the level appearing at the receivers, both from the mobiles and from the transmitters, is below the critical level at which the effect is produced. This in turn necessarily increases the cost of such systems due to the use of two sites, whilst the telephone lines or radio links coupling the sites (and possibly the control also) can put the overall outlay at too high a level.

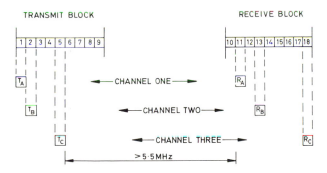

Fig. A13.8 Avoidance of a + b–c Third-Order On-Channel Intermodulation

The alternative method, where a few units only are in use on a single site, is to space the channels, both transmitters and receivers, as if they are of one block. Fig. A13.8 shows the method in detail and indicates an example based on three talkthrough systems on one site. The sequence shown is based on "6 from 18" third order on-channel intermodulation freedom as shown in Appendix 6 (complete channel numbering shown for clarity). Elimination of both 2a–b and a + b – c on-channel products is achieved by the irregular spacing in either and both blocks. Many other combinations are available and can be specified to suit most channel arrrangements. The sequence shown also eliminates radiated on-channel third order inter-modulation (2a–b). The transmit/receive spacing of $>5{\cdot}5$ MHz is included to avoid IF/2 interference. This immunity could however be achieved on a lightly loaded site by limiting Tc – Ra to $>4{\cdot}5$ MHz and Ta – Rc to $<5{\cdot}2$ MHz, thus avoiding $5{\cdot}35$ MHz (i.e. $10{\cdot}7/2$) by $0{\cdot}15$ MHz and still retaining adequate Tx/Rx spacing at VHF.

Appendix 14. Problems Experienced When Mixing 12·5, 25 and 50 kHz Equipment in VHF or UHF Systems

1. General

With the general movement throughout the world towards closer channel spacings in the VHF and UHF bands, there inevitably arrives a time when one or more systems in a given area can include equipments with incompatible channelling.

It is the purpose of this Appendix to point out the possible differences and to suggest ways by which these differences can be minimised whilst the changeover is occurring.

2. Modulation

The type of modulation employed throughout the world fall into two categories, AM (amplitude modulation) and FM/PM (frequency/phase modulation). The two types differ in that AM (assuming a constant percentage depth of modulation) requires a bandwidth proportional to the modulating frequency, whilst FM/PM (assuming a constant modulating frequency) requires a bandwidth proportional to the modulation depth (or deviation, as it is normally called).

2.1 FM/PM. With frequency/phase modulation, the deviation permissible with any given system bandwidth is proportional to the channelling employed, and therefore the change from, say, 50 kHz to 25 kHz channelling necessarily implies that the deviation be reduced by at least a factor of two. This is equivalent to the AM case of reducing the modulation depth from 100% to 50% and obviously will produce a corresponding reduction in the signal-to-noise ratio.

Certain losses can be recovered by improving the signal-to-noise ratio by methods such as reducing audio bandwidth but, in general, the loss is noticeable both in signal quality and (very slightly) in range.

The loss is, however, relatively small when considering the advantages gained by virtually doubling the channel availability.

2.2   AM. In theory, with certain assumptions, it should be possible to reduce the channelling requirements with AM to twice the highest modulating frequency component. This assumes:

(*a*)   distortionless modulation process up to the maximum depth.

(*b*)   no frequency above the highest basic frequency is applied to the modulator. This obviously will limit the general characteristics of the audio source and a circuit may be needed to remove any harmonics.

(*c*)   the carrier frequency stability to be of a very high order.

The above would not be practical and the present state-of-the-art indicates that the channel spacing at VHF and UHF necessary to achieve freedom from interference would approach five times the highest modulating frequency.

Nevertheless, with careful design, it can be seen that the radiated intelligence from a VHF AM radiotelephone system remains relatively of the same level, whether 50, 25 or 12·5 kHz channel spacing is involved (assuming, of course, the same basic modulation characteristics such as response and limiting are maintained).

3.   Selectivity

Obviously, to achieve the desired channel spacing, it is essential that the receiver selectivity is adequate to protect against adjacent channel signals of normal level. It is not the intention of these notes to discuss the effects of high level signals on adjacent channels; this results in a different overall effect and therefore must be tackled in a different manner.

The selectivity of the IF amplifier chain must be adequate to reduce adjacent channel signals to a level which can be tolerated, and the Electronic Industries Association of U.S.A. specify an adjacent channel selectivity of not less than 70 dB. This figure is generally accepted.

The IF filter characteristics must therefore be adequate to fulfil this requirement whilst retaining a reasonably flat "top" to avoid undue changes in other characteristics.

Obviously, the wider the flat top and the steeper the "skirts", the greater will be the excursion possible from the centre point before undesirable effects begin to appear. Unfortunately, this idealised filter is expensive and tends to be larger than desirable and, therefore, a compromise is usually made.

4.   Stability

Stability and selectivity are closely related.

It is essential that the overall drift of both the sending and receiving equipments over the total temperature range specified, plus any drift due to voltage changes, etc., is such that the total excursion of the carrier, including sidebands, etc., does not extend beyond the limits of the filter top response.

In practice, a slight deterioration caused by one sideband beginning to coincide with the start of the skirt can be tolerated but should only be considered when all variations combine in one direction (the unusual case).

Netting is of course very important in a narrow channelling system in order to obtain maximum available space for the tolerances given above.

Fig. A14.1 illustrates drifting of the transmitter and/or receiver from the centre of the filter response. The shaded region embraces the carrier frequency, f_c, and the upper and lower sideband limits f_{s1} and f_{s2} when netted in the centre. The allowable drift can be found by observing the two limits at which sidebands approach the edges of the filter "top". These are the negative drift to f_{c-} and the positive drift to f_{c+}, giving the total permissible drift equal to f_{c+} minus f_{c-}.

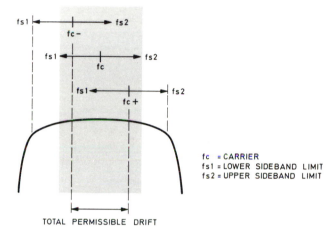

fc = CARRIER
fs1 = LOWER SIDEBAND LIMIT
fs2 = UPPER SIDEBAND LIMIT

TOTAL PERMISSIBLE DRIFT

Fig. A14.1 Limits necessary when considering stability factors

Fig. A14.2 shows the constraints when the sideband deviation extends practically to the edge of the filter "top" response, even when the carrier, f_c, is netted centrally. The diagram indicates that, if drifts as shown before were imposed, attenuation of one or other sideband would occur at high deviation levels. Consequently, the maximum permissible drift, avoiding these effects, is shown to be far smaller.

5. Systems With Mixed Equipments

During the process of changeover, when equipments are being modified to the narrower channelling requirements, some compromise must be made if a system is to remain operational. Additionally, some interference problems are likely if systems on adjacent channels are operating on the wider channelling in the vicinity. Various adjustments can, however, be made to minimise certain of the effects.

5.1 FM/PM Case

(a) If, in one or both directions (fixed to mobile or mobile to fixed) *any* receiver is modified for narrower channelling operation, it is essential that the transmitter deviations in that direction are reduced to suit the new channelling.

Failure to do this will result in distortion in the modified receiver(s).

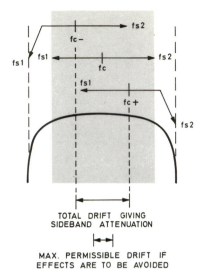

Fig. A14.2 Effects of drift where deviation normally extends
to limit of filter "top" response

Furthermore, if an adjacent system has been modified to a narrow channel operation and is located one channel away at the new spacing, failure to reduce the transmitter deviation as above will result in excessive sideband radiation and, of course, interference into the adjacent system receivers.

Additionally, where the manufacturers specify a tighter tolerance of crystal, this should be fitted to avoid the possibility of excessive drift and of resultant interference to equipments on the adjacent channel.

(*b*)   Where the deviation is reduced, it will be noted that the corresponding receivers (whether modified or not) will require the audio gain control to be advanced to overcome the loss in modulation deviation and consequently the loss in audio output.

(*c*)   When narrow channelling has been introduced, receivers which are still operating in the wideband mode may suffer some interference from an adjacent system using the new spacing and operating on reduced channelling. This is inevitable if the signal level of the adjacent transmitter is high and is a sound reason for the synchronised modification of adjacent systems, if at all possible.

(*d*)   In all cases, accurate re-netting is essential.

Fig. A14.3 shows the results of reducing the receiver bandwidth. With the original receiver filter (upper curve), the deviation is well within the "top" response, giving some latitude for drift. When a narrower filter (lower curve) is imposed, the deviation (to f_{s1} and f_{s2}) extends beyond its top and distortion is produced. Drift can no longer be accommodated, and the deviation must be reduced to the point shown by the asterisks.

5.2   AM Case

(*a*)   As mentioned previously, provided the modulator circuits are arranged to

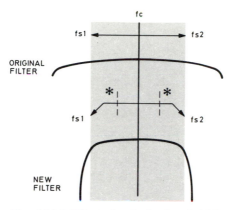

Fig. A14.3 Effect of reducing bandwidth whilst
retaining same deviation

prevent over-modulation and the frequency response follows normal communication practice, then apart from introducing extra crystal stability, nothing more is needed at the transmitters. Some types of transmitters will, however, require slight modifications, usually to clipping levels and these details are usually given by the manufacturer in the modification instructions.

(b)   There should be negligible loss of audio output at the receiver, when changing from a wide to a narrow band system in an AM system.

(c)   Nearby systems using adjacent channels will suffer in a similar manner to 5.1 (*c*).

(d)   As in the FM/PM case, accurate re-netting should be undertaken; but the distortion due to a partial sideband loss will be less than with FM/PM.

Appendix 15. Frequency Control in Radio Communication Systems

1. Introduction

One of the most important requirements in VHF and UHF mobile radio transmitters and receivers is adequate frequency stability.

The stability needed varies inversely with the channel bandwidth, and consequently, over the years, the change from 100 kHz to 12·5 kHz channel spacing has shown that much tighter tolerances over wide temperature ranges are required to ensure reliable communication.

The correct choice of crystals and maintenance of the correct frequency require an understanding of the characteristics of crystals and associated oscillators.

This appendix is intended as a guide to the basic characteristics of frequency control, to assist the engineers responsible for the maintenance of narrow band radio systems.

2. Crystal Holder Styles

Crystal units are available in a number of shapes and sizes, most of which are internationally standardised. Fig. A15.1 shows the more popular styles used by the manufacturers of mobile radio equipment.

It will be seen that the holder styles fall into three main categories: solder sealed, all glass and cold weld. These variations will be discussed more fully later in the section on ageing (5).

3. Specifications

To ensure that a quartz crystal meets the requirements of a particular equipment, a number of parameters must be defined and supplied to the crystal manufacturer.

HOLDER STYLES		INTER-SERVICES	I.E.C.	U.S. MILITARY
SOLDER SEALED	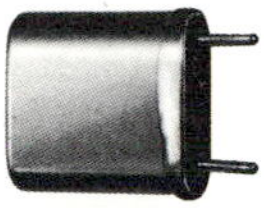	D	AA	HC6/U
		K	CX	HC25/U
		J	BC	HC18/U
ALL GLASS	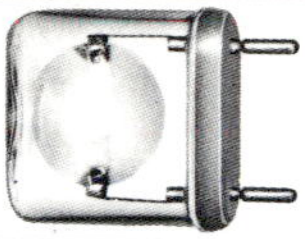	L	DA	HC27/U
	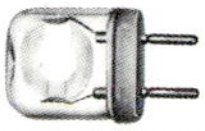	N	CZ	HC29/U
		M	CY	HC26/U
COLD WELD	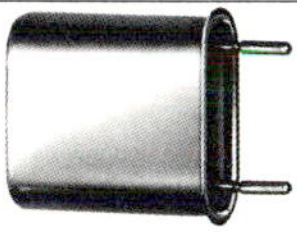	—	DN	HC36/U
	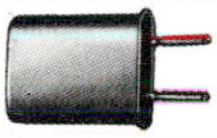	—	DQ	HC42/U
		—	DP	HC43/U

Fig. A15.1 Popular Styles of Quartz Crystal Holders

3.1 Typical Specifications.

Crystal Specifications (Fundamental) No. T40 – Metal Can

1	Holder Style D to DEF 5271A	American equivalent HC6/U
2	Frequency range	3·6–15 MHz
3(a)	Frequency tolerance	To be within +5 to –35 ppm of nominal frequency, measured at 25°C
3(b)	Frequency drift	To be within ±5 ppm over a temperature range of –10°C to +55°C relative to the frequency at 25°C
4	Circuit details	Type: Colpitts. Nominal shunt capacity 30 pF

5	Operating mode	Fundamental parallel resonance
6	Drive level	5 V
7	Activity (to be measured at 30 pF)	To be within the requirement of DEF 5271A
8	Spurious oscillations	To be within the requirement of DEF 5271A
9	Static capacitance	
10	Labelling (impression marking)	(a) Xtal frequency in kHz to 2 decimal places on the top of the holder (b) Xtal No. T40 (c) Serial No. or date code optional

3.2 Frequency Tolerance over a given Temperature Range. Almost all crystals used in communication equipments are of a type described as "AT cut". This specifies the angle at which the basic quartz blank is cut relative to the quartz block axes. If a graph is plotted showing frequency against temperature for an AT cut crystal, the curve is always of a cubic form as shown in Fig. A15.2.

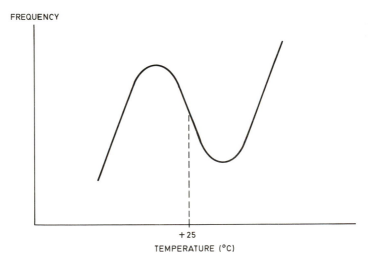

Fig. A15.2 Temperature/Frequency Curve of AT cut Crystal

The actual shape of the curve depends on the exact angle of the cut and the manufacturer normally attemps to hold this angle to within one minute.

Although, in equipment manufacture and maintenance, frequency error is measured in hertz or kilohertz, in crystal specifications the term parts per million is used. Thus, if at a certain temperature, a crystal at 10 MHz exhibits an error of 100 Hz, this should be interpreted as 10 parts per million (10 ppm) to avoid the task of having to calculate at every stage. If we now insert values in Fig. A15.2 for a typical commercial crystal, the relationship between temperature and frequency should appear as in Fig. A15.3. All crystals of the same type, in this case a particular manufacturer's series number, T40, will tend to be the same regardless of frequency.

It should be noted that, relative to 25°C, the curve is symmetrical in shape but not in direction. A typical manufacturer's specification falls into three main temperature ranges:

"T" specification: –10 to +55°C
"E" specification: –30 to +70°C
"P" specification: –30 to +80°C

The best stability that can be achieved economically without temperature control or compensation over these temperature ranges is:

"T" series: ±5 ppm
"E" and "P" series: ±10 ppm

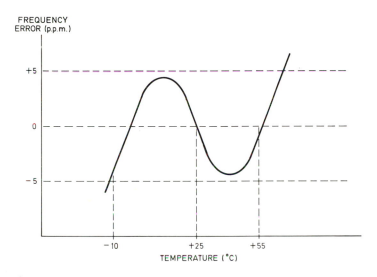

Fig. A15.3 Typical Values of Frequency Variation with Temperature

4. Calibration Tolerances

In the same way in which a mechanical design engineer must allow for manufacturing tolerances by specifying a particular dimension to be within, say, plus or minus 0·1 mm, so must the crystal manufacturer be allowed to include a tolerance on the absolute frequency of a crystal. Not unnaturally, the tighter the tolerance, the higher the cost, so the equipment design engineer must specify the widest tolerance his circuit can accept.

In order to eliminate calibration tolerance from system end-to-end frequency tolerances, reputable communication equipments are fitted with crystal trimming components that will enable the crystal frequency to be adjusted. One exception where trimming is not employed is in the second oscillator circuit of receivers (usually 11·155 MHz). The reason for this is because frequency error in this particular oscillator only makes a very small contribution to the overall frequency errors in the system.

The amount that a crystal frequency can be pulled depends on several factors which need not be discussed here. However, it should be noted that, owing to the extremely small space available in personal portable equipment, the physical size of the trimming component is reduced, and consequently the amount of adjustment to the frequency is less than that possible in mobile and base station equipment. Typically, the latter can be pulled over a total of 60 ppm whereas in portable equipment the figure is normally half that. Consequently, the permitted calibration tolerance in personal portable equipment is less than in other equipment.

5. Ageing

No crystal will remain permanently at the frequency for which it was manufactured. It will gradually change its frequency with time, and this characteristic is called ageing.

Ageing can be attributed to several causes, most of which are known but some not fully understood. Microscopic bodies deposited on the quartz will lower the frequency. Consequently the crystal manufacturer tries to ensure absolute cleanliness in the manufacturing process and makes sure that the quartz is hermetically sealed in its holder. The majority of crystal units produced by the crystal industry are solder sealed; in this process the can is soldered on to the base. Since flux is necessary to ensure soldered joint, flux vapour can condense inside the can after sealing and in time transfer to the quartz and change the frequency. Much time and effort has been spent by crystal manufacturers in developing the solder process to reduce internal contamination, but the solder sealed unit usually cannot be relied on to produce a better ageing rate than 10 ppm in 6 months. However, this figure can be very variable and exceptionally the ageing may be as low as 1 ppm in 6 months.

Ageing rate is not consistent with time, but with a satisfactory hermetic seal will follow the typical curve shown in Fig. A15.4.

Actual values on the curve shown in Fig. A15.4 will vary with individual units, but in general, after eighteen months, the ageing rate will be down to an insignificant level.

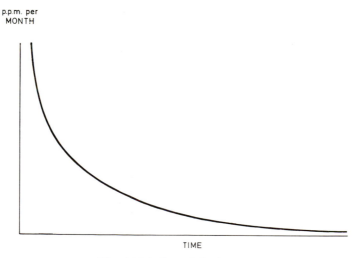

Fig. A15.4 Crystal Ageing

Theoretically the ageing curve can be considered linear on a logarithmic basis, which means that if, for instance, a crystal ages 10 ppm in six months it will age another 10 ppm in the next sixty months, and another 10 ppm in the next six hundred months and so on.

With the improved cleanliness of the sealing process, a glass holder will achieve an ageing rate of about 1 ppm in 6 months – ten times better than that of the solder sealed holders. Unfortunately the production process is more expensive and this accounts for the glass holder being used only where it is essential, for instance in base stations with 12·5 kHz channelling.

Cold weld metal holders are now becoming available in production quantities. These holders have the advantage of the ruggedness of a metal holder coupled with a good hermetic seal, which is achieved by mounting the can on the base under very high physical pressure so that the two surfaces fuse together. Cold weld units are reputed to achieve the low ageing rate of all glass units, and tests so far indicate that this may be true. However, there is no substitute for time and it will take several years of field experience to establish typical ageing rates.

Potentially, cold weld units can be produced at the same cost as solder cans, but this does not include the higher capital costs of the tooling and the cost of the more expensive can and base, both of which must be chemically clean and of higher quality than their solder seal equivalents.

The quality of the hermetic seal on any of the holder styles has a marked effect on ageing since any ingress of air will carry with it water vapour and minute solids. When water vapour condenses on the quartz it will not only change the frequency but may even stop the crystal oscillating at certain temperatures. Normally, crystal units are filled with dry air or nitrogen, but sometimes in the case of glass holders a partial vacuum is established. A trace of helium is often introduced by some manufacturers in order to assist in leakage tests.

Quite apart from the crystal unit itself however, the components of the oscillator circuit will age, but with good design this can be kept down to a minimum.

Changes in the crystal frequency caused by ageing can be in either direction. A crystal unit may often start to change in one direction and then transfer to the opposite direction. As a very general indication however, glass holder crystals tend to increase in frequency with time, while solder seal units tend to decrease in frequency with time.

6. Drive Level

In all oscillator circuits used in most modern communications equipment, only a small amount of power is allowed to dissipate in the crystal unit. However, the level of power may vary from a few hundred microwatts to over one milliwatt, and, although this is mainly of interest to the design engineer, the user should be aware of its implications. The primary effect of power dissipation in a crystal is to increase its temperature; there are however secondary effects. Increasing the temperature, as we have seen in Fig. A15.2, has the effect of changing the frequency.

The crystal manufacturer has therefore to know the circuit drive levels in order to set the frequency within the specified limits. There can, therefore, be a problem in using crystals in circuits for which they are not designed or specified.

Summarising, a crystal may not oscillate on exactly the frequency marked on the holder when plugged into the socket (indeed it is fortuitous if it does) but, provided the crystal manufacturer has been given the correct specification, it should always be possible to adjust the crystal trimmer to the correct frequency.

7. Maintenance of Frequency in the Field

7.1 Effect of Frequency Error. In a perfect radio communication system, all the frequency determining circuits (transmitter oscillator, receiver local oscillators, receiver crystal filter and, in the FM receiver, the discriminator) would be exactly on their correct frequencies and optimum communication would be obtained. However, as it has been shown, it is unlikely that this ideal condition will exist, other than for short periods of time after critical adjustment. However, it is essential that the frequencies are maintained within the specified limits allowed by the equipment designer.

A signal received from a modulated transmitter (either AM or FM) consists of a carrier wave and a family of symmetrical sidebands. This signal can be shown diagrammatically as in Fig. A15.5.

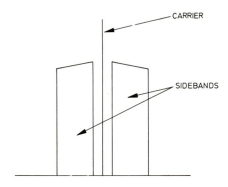

Fig. A15.5 Received Signal Spectrum

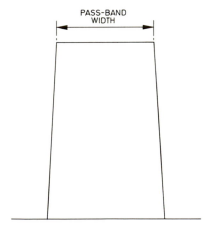

Fig. A15.6 Ideal Receiver Selectivity

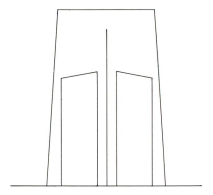

Fig. A15.7 Ideal Case Showing Sidebands Located Within IF Response

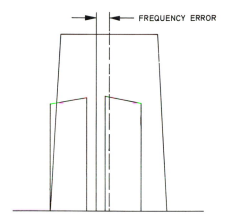

Fig. A15.8 Mis-tuned Case

Although AM and FM are different in detail, Fig. A15.5 can broadly be considered to cover both cases, particularly when considering narrow channelling. The receiver selectivity must be suitable to accept the complete transmitted signal and ideally the selectivity curve would appear as in Fig. A15.6.

It would then pass the transmitted signal without modification as in Fig. A15.7.

If, however, there is a frequency error in the system, the transmitted signal will not sit centrally within the receiver selectivity curve (Fig. A15.8) and it will be seen that one transmitted sideband will be partly cut off by the receiver selectivity. The effect of this will be that the audio signal heard from the receiver loudspeaker will be distorted. One effect, peculiar to FM systems is that impulse noise interference (particularly ignition noise) becomes rapidly worse as system frequency error increases.

Frequency changes within a communication system are usually very gradual, and distortion increases gradually with frequency error. Therefore the system will be degraded very gradually with time owing to crystal ageing, and thus the system will need regular frequency maintenance. Since ageing rate reduces with time, frequency maintenance on straightforward systems will be needed probably four times during the first year after installation and thereafter twice a year.

The amount of frequency error that can be tolerated is dependent on channel spacing. For 12·5 kHz channel spacing an error of 1 kHz is the maximum permissible, whilst for 20 kHz and 25 kHz channelling the figure is 1·5 kHz. Some countries may have licensing regulations which require tighter limits.

Frequency errors in excess of those quoted above give rise to severe distortion, high levels of impulse interference, lack of intelligibility and loss of range. AM systems tend to be more tolerant to frequency error than FM systems.

7.2   Measurement of Frequency

7.2.1   *Test Equipment.* If the absolute frequency of the system is not of prime importance (for instance, where the adjacent channels are in use some considerable distance away), good absolute frequency measurement is not essential except to satisfy licensing requirements. In these cases it is usual to assume that the base station frequencies are correct and adjust all mobile transmitter and receiver frequencies to remove end-to-end error. Marker oscillators have been developed to assist in this adjustment. They are available for use at the various intermediate frequencies used in mobile radio equipment, for example,

10·7 MHz
455 kHz
23·455 MHz

The marker oscillator generates a signal in the centre of the intermediate frequency pass band. When a signal is received from a transmitter, the receiver will produce a signal at IF and if, at the same time, the relevant marker oscillator is coupled up to the receiver IF circuits, a beat note will be heard at the receiver output. The frequency of the beat note will be equal to the end-to-end frequency error of the system.

If a mobile receiver frequency is being checked, then end-to-end frequency error will be eliminated by adjusting the receiver crystal trimmer to produce zero beat. This process is generally termed netting.

If, however, the mobile transmitter frequency is being checked, although it can be measured by the same method, adjustment becomes a two man job since the marker oscillator must be used at the base station receiver. One person at the base receiver passes instructions to the other at the mobile transmitter.

The marker oscillator is also used to net in a signal generator to a receiver, for instance, when an FM discriminator is being adjusted.

Where an accurate frequency measuring equipment or counter is available, the task of adjusting transmitter frequencies is simplified.

The following points should be noted:

(i)   Since the transmitter usually consists of a crystal oscillator followed by a number of multipliers, any frequency error of the crystal oscillator will also be multiplied and, therefore, it is always best to measure the output frequency of the transmitter in preference to the crystal frequency.

(ii)   Apart from the improved accuracy, measuring at carrier frequency can be carried out without opening the equipment and high levels are normally available.

(iii)   The levels at the crystal oscillator are invariably very low and often not enough to operate the counter; attempts to couple in the counter often result in the "pulling" of the crystal so that when the counter is removed the frequency will change.

(iv) Counter measurement of receiver frequencies is best achieved by measuring the output of the oscillator multiplier chain. At this point a reasonable level exists and should be adequately buffered from the crystal oscillator. The frequency here differs from the wanted receive frequency by plus or minus the first intermediate frequency.

Unfortunately, counters have a psychological effect on many engineers in that, as the frequency is displayed on a digital basis, the reading is considered accurate and beyond question. However, the internal frequency standard in the counter suffers from the effects of temperature, ageing and setting errors in exactly the same way as the crystals in the equipment under test. Thus, the counter oscillator needs to be checked for accuracy on a regular basis in order to achieve accurate measurements. Most high quality counters have oscillators which have been specially developed to minimise ageing and temperature effects, but checking against a standard frequency is essential, at least every three months. It is important to realise that the counter "self check facility" checks only the logic circuits and not the accuracy of the internal crystal oscillator.

It is recognised that any measuring equipment, whether it be a micrometer, weighing machine or frequency counter, should be capable of an accuracy at least ten times better than that of the desired limits of measurement. If, for instance, a frequency needs to be set to within 1 ppm, the counter should possess an accuracy of at least 0·1 ppm.

Returning to the maximum end-to-end frequency errors allowable over a mobile radio system, i.e. 1 kHz at 12·5 kHz channelling and 1·5 kHz at 25 kHz channelling, the errors translate into approximately 6 ppm and 9 ppm respectively in the 160 MHz band. In the 450 MHz band, the errors become 2 ppm and 3 ppm respectively. Counters used in frequency checking should therefore have an accuracy of not worse than 0·2 ppm if used at UHF, and better than 1 ppm at VHF.

Counters are normally checked against one of a number of standard frequency transmissions available throughout the world – WWV in the USA and MSF in the UK are examples. A communication receiver should be used to receive the standard frequency transmission and the internal crystal oscillator of the counter is also coupled to the receiver in such a way that a beat note is heard. The counter oscillator should then be adjusted for zero beat. The adjustment is best undertaken at the highest possible frequency.

7.2.2 *Effects of Ambient Temperature.* In Section 3.2 it is shown that the curve relating frequency and temperature is symmetrical (but not in direction) relative to 25°C. Thus, if the specification calls for ± 10 ppm, the tolerance is relative to the frequency at 25°C.

In other words, over the temperature range of, say –20°C to +60°C, the frequency will remain within 10 ppm of the frequency at 25°C. The frequency may shift either low or high depending on temperature, but never more than 10 ppm. The difference between the maximum frequency and minimum frequency may, however, be as high as 20 ppm. See Fig. A15.9.

If the crystal is pulled to the correct frequency at +25°C (point A), the maximum error of nearly 10 ppm over the temperature range will be at points B & C. However, if it is pulled to the correct frequency at the temperature corresponding to point B, the maximum error will occur at only one point (point C) and the frequency will be approximately 20 ppm high. Conversely, setting at point C will result in a maximum error of 20 ppm low at point B.

It is now obvious that all frequency maintenance adjustments should ideally be at +25°C if the equipment is likely to be subjected to the full range of ambient temperature (−10°C to +55°C in the above example). This immediately creates a field problem, since the maintenance engineer often cannot make any temperature changes at the time of the frequency adjustment. He can attempt some corrective measures in a tropical country by making adjustments in an air-conditioned room, whilst in cold climates raising the room temperature by a fan heater is often sufficient.

The problem is generally peculiar to temperate climates where wide temperature extremes are likely. In tropical zones adjustments at point B in Fig. A15.9 will normally be adequate since the temperature corresponding to point C may never occur.

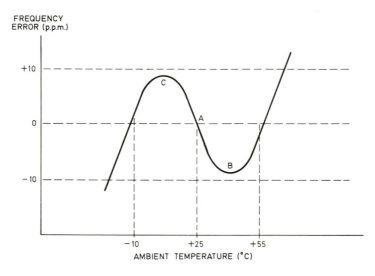

Fig. A15.9 Frequency Error Relative to Temperature and Adjustment

8. Crystals in Ovens with Thermostat Control

To achieve almost complete isolation from external temperature effects, a crystal can be placed in a thermostatically controlled oven, the temperature of which can be set to a value higher than would normally be encountered under operational conditions. In many equipments such an oven would be set to +80°C. Most ovens used will hold the temperature within 1°C at any one ambient temperature within the range −30°C to +60°C, but the average internal temperature over the total range may vary up to ±3°C. Thus, if the oven is set to +80°C at +25°C, at −30°C the average temperature could be +77°C and at +60°C it could be +83°C. The crystal unit will thus be held within $\pm 3°$C over the complete ambient temperature range.

In addition to this there is a manufacturing tolerance of $\pm 2°$C on the nominal $+80°$C setting. Consequently the crystal specification will allow for this error by calling for a ± 5 ppm maximum frequency change over the maximum oven internal temperature range. Typically however, in modern equipment, the frequency stability achieved with oven control is ± 2 ppm over the temperature range $-30°$C to $+60°$C.

Since the oven temperature is continually cycling $\pm 1°$C, the frequency will also be cyclic, and in maintenance, adjustment will be necessary to judge the point about which the frequency cycles.

Another factor which should be noted is that ageing is accelerated by heat and consequently, adjustments may be needed at short intervals initially, although the required order of stability may be reached in a shorter time.

Crystal ovens used by reputable manufacturers have a snap action thermostat for optimum reliability but nevertheless have a life limited to probably three or four years. Because of limited life and because the 6 watts consumption is often more than that taken by a complete equipment without ovens, they are no longer used except when considered essential.

9. Temperature Controlled Crystal Oscillators

In order to reduce the problems of life and power requirements of a crystal oven, yet still achieve good stability over a wide temperature range, another system of heating the crystal has been developed. In this system the crystal unit is mounted in a block of aluminium which is heated by a power transistor. A temperature sensitive resistor is employed to feed a control voltage to a DC amplifier which, in turn, controls the current flowing through the power transistor. By this method close control of temperature is achieved; temperature cycling and hence frequency cycling is eliminated. These oscillators are now being used in high class base stations where a frequency stability of better than 2 ppm over the full temperature range $-30°$C to $+60°$C is necessary for either licensing or system requirements.

A precision version of this type of oscillator has been used in transmitters where a frequency stability of $0·1$ ppm over the temperature range $-30°$C to $+70°$C is necessary in quasi-synchronous FM systems. An extremely high stability drive source is used.

In AM quasi-synchronous systems it is necessary to maintain frequencies to an accuracy of $0·01$ ppm. The only practical way of achieving this standard is to generate the required frequency from a "standard" oscillator on 1 or 5 MHz. One well known manufacturer uses a precision 5 MHz oscillator as a reference frequency to which the output frequency is locked by synthesiser techniques. The basic performance of the oscillator is a frequency stability of $0·004$ ppm over the temperature range $-10°$C to $+60°$C and almost more important, an ageing rate several orders better than that expected from a conventional oscillator.

10. Temperature Compensated Crystal Oscillator

It is possible to design electrical circuits which tend to straighten out the curve of Fig. A15.2 into a horizontal line. An attraction of this approach is that the compensation circuit consumes minimal power and takes up very little space. The main drawback, however, is that, for optimum results, the compensation circuit has to be individually tailored to each crystal. It is therefore costly and creates a replacement problem in the event of failure. A compromise solution is to design a small number of compensation variations and match crystals to these variations. Owing to problems of specifying the crystal characteristics when obtaining crystals from more than one supplier and the lack of availability of these special crystals in quantity, temperature compensation is not yet used as a general solution to low consumption frequency stability over a wide temperature range.

With two stage compensation, the performance of the crystal oscillator can be of the order of ± 5 ppm over the temperature range $-30°C$ to $+60°C$.

11. Sleeping Sickness

"Sleeping sickness" is a somewhat unscientific term applied to a peculiar and little understood characteristic of crystals. Fortunately it is quite rare; but it needs to be mentioned.

Some crystals, which under normal manufacturing tests are perfectly satisfactory, fail to oscillate after a period of inactivity and this may happen when an equipment is manufactured, fitted with its crystal and dispatched after tests to a customer.

If the period between factory dispatch and customer receipt is lengthy, the customer finds that the equipment does not work, apparently because a crystal is faulty when in fact it is suffering from "sleeping sickness". He returns the crystal for replacement and it is checked in a crystal test set. The tester finds nothing wrong with the crystal; all tests indicate that it is perfectly satisfactory. The reason for this paradox is that in order to achieve good frequency stability, the crystal oscillator in the equipment drives the crystal at a low level, whilst the crystal test set tends to drive the crystal somewhat harder.

"Sleeping sickness" in a crystal can be temporarily overcome by electrically shocking the crystal into activity. If the equipment is used regularly the effect does not occur, but the crystal will "go to sleep" again if not activated for a period. To continue the story, the crystal is returned to the customer, and on the way the crystal "goes to sleep" again. The customer considers he has a faulty crystal; the equipment manufacturer considers the crystal is good . . . !

The onset of "sleeping sickness" usually occurs more than six weeks after manufacture.

12. Maintenance of Frequency Stability in Quasi-synchronous Systems

In quasi-synchronous systems, both AM and FM, a very high stability frequency source is used. In a particular high grade frequency generator the frequency is controlled by a 5 MHz precision oscillator to which a conventional crystal oscillator (the frequency of which is determined by the wanted carrier frequency) is locked using techniques similar to those used in digital frequency synthesisers. The 5 MHz precision oscillator has an accuracy and stability far better than almost all frequency measuring equipment used in the field. Thus there arises a maintenance problem. In N.W. Europe the Droitwich transmitter can be used as a frequency standard using a low cost "off-air standard receiver". This type of receiver, which is available from several suppliers, receives the 200 kHz Droitwich transmission, removes the amplitude modulation and multiplies the frequency by five to give a highly stable 1 MHz output. This 1 MHz output can be used as the standard for a frequency counter. Most counters have provision for an external 1 MHz input. A similar receiver for standard frequency transmissions elsewhere in the world may be available. The counter used in these measurements must have the facility to display to 0·1 Hz.

Alternatively, since the frequency difference between transmitters is of major importance, it is suggested that measurements of this difference could be the method adopted. In practice, the measurements can be tedious in that the first requirement is for an AM receiver having a frequency response extending down to at least 30 Hz. This receiver must be sited in a position where it will receive approximately equal signals from two transmitter sites.

The output from the receiver is then fed into a frequency counter to establish the frequency difference. If frequency adjustment is necessary, the engineer measuring the difference frequency must pass instructions to a communication engineer at one of the transmitter sites as to the amount of adjustment necessary. This method has two advantages. Firstly it measures directly the frequency difference, and secondly the counter used can be quite crude in terms of frequency stability since 1 Hz error in, say 100 Hz is relatively unimportant and represents 10,000 ppm. For frequency differences of less than 2 or 3 Hz, the slow beat can be counted and the difference frequency calculated with the aid of a wrist watch.

Appendix 16.   Radio Communications in Confined Spaces

1.   Introduction

The use of radio as a method by which intelligence may be passed from one point to another without the use of intervening physical conductors is, of course, well known. Its use over long distances has been accepted for a considerable time, whilst its ability to be used to and from vehicles in defined areas is typical of many private, public and governmental communication systems in operation throughout the world.

It is only recently however, with the introduction of personal portable equipment, that the obvious need to communicate within very limited areas has arisen to any great extent. These areas are often very confined and, in many cases, heavily screened and therefore special techniques are necessary to ensure adequate coverage.

The following paragraphs are intended to illustrate the various applications and to show possible ways of solving the communication problems in the situations given.

2.   Types of Application

There are several applications:

    (a)   *Tunnels*
        Types of tunnel include:
        1.   Railway – brick lining, including cut and cover.
        2.   Railway – steel tubing.
        3.   Railway – brick walls and ceiling, earth floor.
        4.   Mine – rough hewn and propped.
        5.   Sewers – various.

    (b)   *Shafts*
        1.   Mine shafts – with lift gear etc.

(*c*)   *Buildings*
1.   High rise steel structures.
2.   Underground buildings.
3.   Shopping precincts, arcades etc. (above and underground).

(*d*)   *Miscellaneous*
1.   Ships' holds, passenger liners etc.
2.   Restricted coverage of roads through cuttings etc.

3.   Principle of Operation

3.1   General. Standard techniques involving antennae as radiators and pick-up devices have several disadvantages when used in confined spaces. First, due to multiple reflections and high absorption losses, the range from an antenna is reduced to a fraction of the range in normal free space. This results in the coverage of, say a tunnel, being but a short distance even with quite high radiated powers.

The second disadvantage is that, being a relatively concentrated source of signal, multiple reflections are extremely marked and, even if the range is considered to be adequate, the points of in-phase and out-of-phase reflections would cause extremely erratic communication, particularly whilst moving. The third disadvantage is the size of the radiator, especially when considered for use in a small tunnel or shaft.

Obviously, therefore, if the transmitted signal (and of course the received signal) could be *released* and *collected* respectively at regular small intervals along a tunnel then the first two difficulties mentioned would be overcome. A point source would not exist and therefore multiple reflections would be minimised. Furthermore if the radiator could be concentrated into a device of a small cross-section then the size problem also would disappear.

Fortunately, this can be achieved quite simply by the use of certain types of transmission line or cable which permit, either normally or by design, a limited *leakage* or pickup of the transmitted signal.

Initially, tests were undertaken and installations carried out using balanced 200 ohm TV type transmission line, but this type of cable required baluns to match 50 ohm unbalanced to 200 ohm balanced and also suffered from increased losses unless mounted clear of nearby surfaces consisting of either metal or lossy insulation. The long term loss increase of balanced 200 ohm cable is also quite high, particularly in contaminated environments, owing to the build up of conducting deposits on the outer surface.

The tendency over the past few years has therefore been to change to coaxial cable, where the outer is substantially at earth potential and the leakage is controlled either by the weaving of the outer mesh or by definite machined slots in a solid copper outer sheath.

The latter method has been generally adopted as the most satisfactory both in radiation consistency and in physical strength. However, regardless of the advantages and disadvantages of the various cables, the operational principles remain the same.

3.2   Cable characteristics. Assuming there is no leakage from a given cable, then if a long length of cable is correctly matched to a signal source, the resultant power at any point along the cable will be equal to the source power less that lost in the attenuation of the cable to that point.

Let us now introduce a controlled amount of leakage by, say, slotting the coaxial outer sheath. From the slots will be radiated a defined amount of energy which, in turn, will extend to a given distance from the cable. The power available at this distance can be derived from the attenuation of the cable (from the transmitter to the slot nearest the receiver) plus the coupling loss from the cable to the receiving point. It should be mentioned that when radiating characteristics are introduced by slotting etc., the attenuation of the cable per unit length increases slightly and therefore the final attenuation figure used for the cable must be that of the *slotted* cable.

It can be seen that the received signal will be determined by the transmitter power, the length of cable involved, and the coupling loss over the desired distance (normally taken as 6·1 m or 20 ft). Therefore it is essential to calculate the various parameters bearing in mind the lowest transmitter power – usually in the direction feeding the base station and possibly associated with portables rather than mobiles.

3.3   Practical parameters. A typical slotted type of radiating cable would normally be approximately 13 mm (0·5 in) in diameter. At 150 MHz, the cable attenuation would be 3·6 dB per 100 metre (328 ft) length, whilst the coupling loss over a 6·1 m (20 ft) distance from the cable would be 75 dB.

Assuming a transmitter of 1W, to achieve a signal at the receiver of 1 μV, a total attenuation of 137 dB could be tolerated. Allowing for the coupling loss of 75 dB, the resulting cable length permitted becomes $(62/3\cdot6) \times 100$ m = 1·72 km or just over one mile.

Coupling loss between the cable and a point of transmission or reception in free space will increase at the rate of 6 dB each time the distance is doubled. However, in a confined area, the increase is likely to be greater with the actual amount being a function of the type of screening, size of surrounding area, etc. The actual figure might well be as high as 20 dB but more likely will be of the order of 10 to 12 dB.

Obviously, therefore, any increase in range from the cable must be off-set by a reduction in cable length. This is not normally serious as any increase in area to be covered is usually accompanied by a considerable reduction in cable length. The coverage of a high rise building is a typical case, where the area is increased but the height is small compared with the length of a tunnel.

Alternatively, if a number of areas require coverage then duplication of the cable fed from a common source is the obvious answer.

3.4   Matching of cable. To ensure maximum power transfer from a transmitter into a given load, the load impedance should equal the output impedance of the transmitter. A similar state exists when considering a receiver/cable combination where the receiver input must load the cable correctly. Where a transmission line is connected to a transmitter, the cable must possess the correct characteristic impedance equal to both the load (antenna) and the source.

However, if the cable is of considerable length with an attenuation of tens of dB, the load becomes reasonably isolated from the transmitter and the value of termination (or lack of it) will not affect the performance of the transmitter by reflections back into the transmitter. They will be minimal and determined solely by the accuracy of match between the transmitter and the characteristic impedance of the cable.

This particular condition exists with radiating cable systems and, provided the length of cable is sufficient for its attenuation to swamp the load condition, any termination (or none) can be used without affecting the transmitter.

Nevertheless an effect will appear at the far end of the cable if the termination is incorrect.

Let us consider the last, say, 50 metres of cable. By correctly terminating the cable the power at the 50 metre point will flow smoothly and without appreciable reflections into the load, and therefore radiation from the last 50 metres will decrease in a linear fashion. If, however, no load is connected, standing waves will exist along the last section of the cable and reflections will flow back into the cable until dissipated by the attenuation.

Along this length there will be distinct maxima and minima of radiation strength and this could prove detrimental to the coverage from the cable. It is therefore recommended that, although of no benefit to the transmitter, the cable should always be terminated at the far end. The power at that point will be quite low so that high power terminating resistors are unnecessary.

4. Two-Direction Operation

It can be seen (Fig. A16.1) that a single length of cable having a defined loss characteristic will at some point in its length, radiate insufficient power to span the distance between the cable and the mobile or portable (coupling loss). Additionally, the limited power of the mobile or portable in the opposite direction will, together with the cable loss, be inadequate to feed the base station receiver.

Obviously, therefore, the point where the system is just operational ("A" in the diagram) indicates the limit of range.

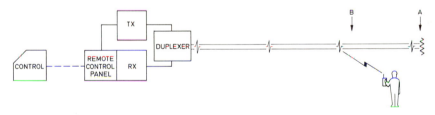

Fig. A16.1 Uni-directional Operation

If the mobile or portable is moved towards the base station, to a point where the signal level at the portable is 3 dB above the level determined for a just-operating condition, then it can be seen that, by halving the base station power, the just operating condition will re-appear in the outward direction at the new location (point B). Depending on the cable loss, this point could be 30–100 metres (100–330 ft) from A and would usually be but a small percentage of the total length, which could be up to 1·7 km for 0·5 in dia. cable at 150 MHz.

It can now be seen (Fig. A16.2) that by diverting the unused half of the power (together with the receiver feed) to a second similar length of cable running in a different direction to the first, the effective length of the total cable run will now be almost twice a single length (i.e. about 3·2 km).

Matching the two cable lengths into the base receiver will obviously reduce the effective sensitivity of the receiver by approximately the same amount as given above, thus producing a similar change in effective length after splitting.

The number of splits in a feeder can obviously be extended as required but it must be borne in mind that the power available in the direction of portable to base station will be the limiting factor. Also affecting the final arrangement will be the matching network configuration at the base station.

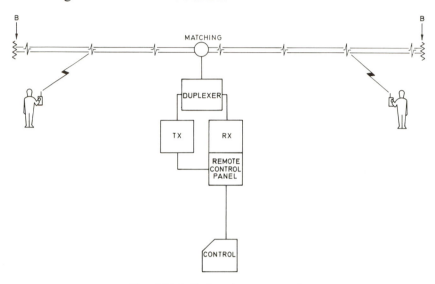

Fig. A16.2 Bi-directional Operation

5.   Additional Equipments

When the coverage from one station ceases to become effective, then it may be necessary to add a further station on the same channel in order to achieve coverage of an additional section of the total area to be spanned. The additional station would tend to follow the principles outlined in Section 4 and would therefore extend the coverage by a similar amount (see Fig. A16.3), giving a total length of about 6 to 6·5 km at 150 MHz with 0·5 in. dia. cable. For talkthrough operation, a two-input controller would be equipped with "connect" facilities.

Control of the station would follow normal remote control techniques. To achieve inter-unit talkthrough, some complication of the control would be necessary to provide the connect facilities needed.

6.   Cable Repeater or Amplifiers

Fig. A16.4 shows a method whereby the output of the transmitter is boosted at intervals along the radiating cables, whilst at the same time (using the same amplifier or amplifiers) the received signal from the mobile or portable is increased to overcome the cable loss between the reception point and the receiver.

DC power can be fed over the cable if desired to the amplifiers whilst the gain of the units should be such as provide stability consistent with reasonable cable lengths between amplifiers. Short lengths of non-radiating cable may be needed at the input

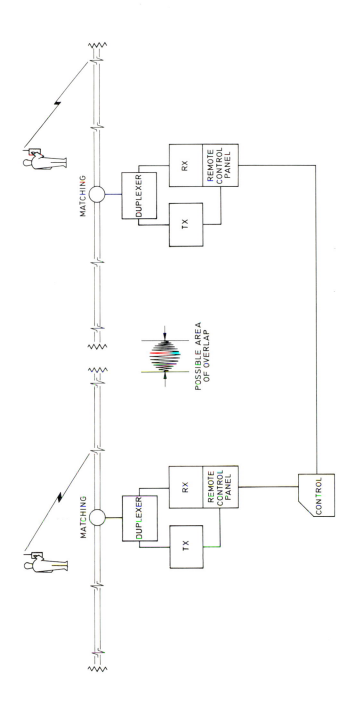

Fig. A16.3 Multi-Station Operation

and output of amplifiers with high gain to provide sufficient isolation against unwanted feedback. Under normal conditions, using amplifiers of average gain, such precautions may, however, be unnecessary.

Multichannel operation is possible with this type of repeater, provided the amplifier has sufficient bandwidth to accommodate the frequencies involved.

Spacing between amplifiers will mainly be a function of amplifier gain and cable loss, but it is also necessary to consider the transmitter output power and receiver sensitivity to maximise the layout efficiency. The transmitter feed point could be at the centre of a two-way split, as with conventional systems. In other words, Fig. A16.4 would be repeated in the opposite direction, provided the 3 dB splitting loss were taken into consideration.

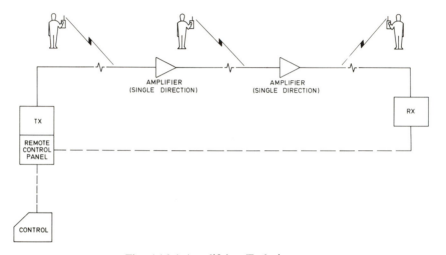

Fig. A16.4 Amplifying Technique

7. Interconnected Equipment

This method has been considered in order to achieve suitable coverage of extended tunnel sections whilst at the same time retaining the ability to operate mobile-to-mobile in a talkthrough mode without complication of the control facilities.

Fig. A16.5 shows two base station transmitter/receiver units designed into a tunnel system to achieve maximum coverage. The total length of cable (at 150 MHz) could be 6 to 6·5 km. The control station operates on reverse frequencies and would be "injected" into the system at a convenient point. The external antenna is shown in order to provide coverage (if desired) outside the tunnel.

The two units are then connected together using screened quad cable and terminating units at each end. The operation of the system is based on both station and link talkthrough with the switching facilities phantomed on the speech pairs.

Thus a signal arriving at receiver Rx 1 will be re-transmitted both by transmitter Tx 1 and transmitter Tx 2, whilst Rx 2 will pass signals for re-transmission to Tx 1 and Tx 2.

Examination of Fig. A16.5 will show the actual AF talkthrough path located at a point midway between the two units. This has been done to ensure similar delay

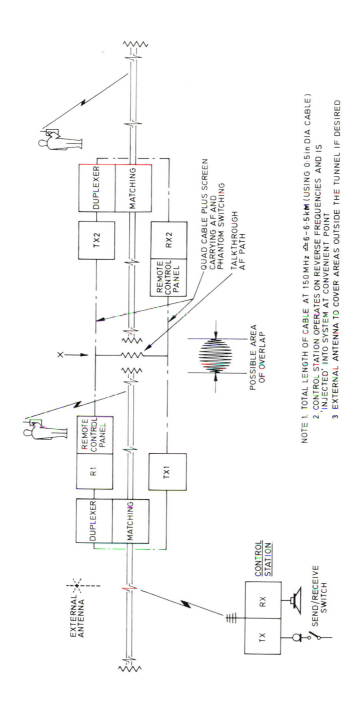

NOTE 1. TOTAL LENGTH OF CABLE AT 150 MHz ≈ 6-6.5 km (USING 0.5 in DIA CABLE)
2. CONTROL STATION OPERATES ON REVERSE FREQUENCIES AND IS 'INJECTED' INTO SYSTEM AT CONVENIENT POINT
3. EXTERNAL ANTENNA TO COVER AREAS OUTSIDE THE TUNNEL IF DESIRED

Fig. A16.5 Possible Repeater Configuration

characteristics of the transmitted AF at the point where the signals could overlap (point X) if the channels employed were identical.

The diagram also shows where the use of a conventional antenna can be included to achieve coverage of an open area.

8.   Coverage of Building & External Areas

Here the method follows closely that of other applications such as tunnels, shafts etc. but uses the radiating cable technique as a means of covering the internal area of the building whilst a conventional antenna system is used for external coverage.

Fig. A16.6 shows a method used to achieve coverage of this type. The radiating cable feeds the antenna which is mounted on the roof of the building whilst at the same time providing coverage inside the building by direct radiation from the cable. The power radiated from the antenna is equal to the transmitter power *less* the power radiated from the cable and that lost through cable attenuation. This method is often employed for paging systems.

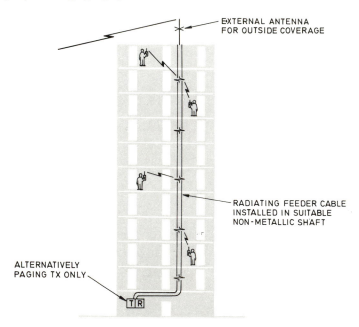

Fig. A16.6 Typical Solution in High Rise Building

The cable is normally laid within the building up or along a convenient shaft, but care must be taken to avoid using metallic shafts where the screening could prevent adequate radiation throughout the area.

9.   Restricted Open Areas

Where restricted coverage is required along, say, a section of motorway which passes through a cutting or tunnel, radiating cable can be used in the same way as with other types of tunnel or shaft. It is possible for the cable to be fed with several simultaneous channels.

10. Miscellaneous Applications

Locations such as ships' interiors and shopping precincts are all likely areas in which radiating cable techniques can be of use. Applications must be considered individually and are therefore not examined in detail in these notes.

11. Frequency Bands

Figs. A16.7 and A16.8 illustrate the performance (attenuation and coupling loss) of a variety of cables as functions of frequency. Here, coupling loss is defined as the average difference between the signal level in the cable and the signal received by a 0 dB gain antenna 20 ft (6·1 m) from the cable.

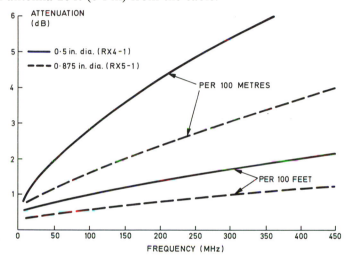

Fig. A16.7 Attenuation as a Function of Frequency for Slotted Cables (Andrews Radiax)

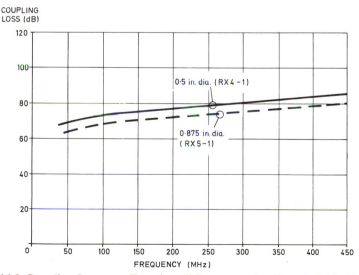

Fig. A16.8 Coupling Loss as a Function of Frequency for Slotted Cable (Andrews Radiax)

The choice of frequency bands depends very much upon the application. Obviously, the lower the frequency the less will be the cable attenuation and, therefore, the longer will be the possible cable span. Additionally, coupling loss will be reduced at the lower frequencies giving a greater effective coverage around a given cable or, alternatively, the possibility of extending the cable length still further. Figures of 1·5 dB per 100 m (328 ft) attenuation are normal for the type of cable mentioned in Section 3.3 and operating at 30 MHz.

However there are disadvantages. At 30 MHz, antenna systems used with portables and mobiles become larger and therefore more unwieldy in confined spaces. Noise pickup is likely to be more objectionable with electrical switching noises predominating.

At the other extreme (450 MHz), the attenuation is approximately twice the 150 MHz figure (7 dB per 100 m (328 ft)) with an increase of 10 dB in the coupling loss. The length of the cable at 450 MHz will therefore be approximately 40% of the 150 MHz equivalent length. On the other hand, portable and mobile antennae are extremely small, electrical noise is low and, additionally, any standing waves caused by reflection are extremely close together and therefore less objectionable.

On the whole, the best compromise, bearing in mind the range and convenience of operation, appears to be the 150 MHz band and it is recommended that due consideration be given to this portion of the spectrum.

12. Frequency Planning

One great advantage of the radiating cable technique over free space radiation concerns the limit of coverage. As the majority of the propagation is over the cable and only a defined amount is available for free space radiation, the actual field at the cable end has a relatively small radius. Thus, if the end of the cable of a second system approaches the end of the cable of the first system, both systems on the same channel, there will be only a limited distance where the two signals are approximately equal at the mobile or portable. In any case, immediately the difference exceeds, say 12 dB the stronger signal will capture.

It can therefore be assumed that, provided the frequencies are within 200–300 Hz of one another, no appreciable interference between two systems on the same channel will exist. In fact between 10 Hz and 150 to 200 Hz quasi-synchronisation conditions will exist.

Above 200 Hz, a beat note may become obvious although not objectionable, whilst the phase delay characteristics should be minimal if suitable precautions are taken. In the case in point, the location of the AF talkthrough path resistor at the centre of the system will tend to equalise the phase delay errors (Section 7).

However, even under the most adverse conditions, it is calculated that a distance of less than 15 metres (50 ft) will be affected. Distances considerably shorter than this figure can often be achieved by suitable adjustment of the cable lengths to reduce the overlap between the ends.

Appendix 17.   Tone Lock

1.   General

Tone lock, or tone squelch as it is often called, is a method of preventing the over-hearing of unwanted transmissions from other users on the channel. Technically, the term "tone squelch" should be used when referring to the receiver unlocking function.

The system is often referred to as "dual squelch" because it uses both tone and carrier signal-to-noise operated squelch circuits in series.

The principle is based upon the simultaneous radiation of a low level sub-audio tone superimposed on the carrier of wanted transmissions. Reception of this tone opens the equivalent tone squelch enabling the wanted signals to be heard. Unwanted transmissions on the same channel cannot be heard by the operator because they do not include the sub-audio tone required to open the tone squelch of the receiver.

Simultaneous reception of a wanted signal and an unwanted signal can, however, occur, resulting in the normal type of interference expected from co-channel operation.

2.   Tone Frequencies Employed

Two groups of frequencies associated with the tone lock systems are allocated by the Electronic Industries Association (E.I.A.). The groups and frequencies are as follows:

Group A

67·0 Hz	151·4 Hz
77·0 Hz	162·2 Hz
88·5 Hz	173·8 Hz
100·0 Hz	186·2 Hz
107·2 Hz	203·5 Hz
114·8 Hz	218·1 Hz
123·0 Hz	233·6 Hz
131·8 Hz	250·3 Hz
141·3 Hz	

Group B

71·9 Hz	146·2 Hz
82·5 Hz	156·7 Hz
94·8 Hz	167·9 Hz
103·5 Hz	179·9 Hz
110·9 Hz	192·8 Hz
118·8 Hz	210·7 Hz
127·3 Hz	225·7 Hz
136·5 Hz	241·8 Hz

The deviation limits permitted for use with any of the above tones in an FM system are: 50 kHz equipments 0·9 to 2 kHz
25 kHz equipments 0·5 to 1 kHz
15 kHz equipments 0·3 to 0·6 kHz

Owing to the low deviation, coupled with the additional filtering and the response characteristics of the audio system, the above tones are sufficiently attenuated to be inaudible to the operator. When choosing tones, those in the centre range of each group are preferred. They are a compromise between the possibility of overhearing any tone leaking through the audio section and the ease of achieving adequate deviation which is in turn governed by the low frequency fall off of the system audio response.

3.   System Limitations

Some basic limitations exist in the use of tone lock.

3.1   Transmission Mode. Tone lock is suitable for transmission over both amplitude modulated and frequency modulated systems.

Until recently, AM systems have been restricted by the difficulty of ensuring an adequate audio response with sufficiently low distortion at the lower frequencies. Modern equipment design has overcome this problem.

3.2   Transmission of Tones over Wire Circuits. The transmission of tones over short cable lengths employing normal telephone pairs is feasible. However, when the telephone pairs include hybrids, isolating transformers, loading coils, relays, etc. they affect the line characteristics below 300 Hz so as to prevent the transmission of sub-audio tones at adequate levels.

Additionally, distortion at low frequencies caused by iron in the components is prohibitive.

Consequently, tone lock is limited to cases where the module can directly modulate the transmitter or be directly fed from the receiver. At the low frequencies involved, the loss is minimal in these applications.

4.   Methods of Use

4.1   Locally Controlled Equipment. In this category, the problems associated with Section 3.2 are eliminated and tone lock can be achieved by simply fitting suitable modules in or adjacent to both the transmitter and receiver units, assuming that tone lock is required in both directions.

The operation of the send/receive key automatically causes the carrier to radiate together with the chosen sub-audio tone and the necessary speech modulation.

In the receiver, the tone squelch module opens the audio path when a matching tone modulated carrier is received but remains closed to other signals not using an identical tone system.

An additional feature of a locally controlled tone lock system is that the tone squelch in the receiver can be inhibited to enable the operator to listen out before transmitting. The procedure avoids unnecessary jamming of signals by indiscriminate transmissions.

4.2   Remotely Controlled System over Line Pairs. In this mode of operation, an arrangement similar to that of Section 4.1 can be used. The tone lock modules can be located in the transmitter and receiver and directly connected into their respective units.

To inhibit the tone squelch and enable the receive half of the channel to be monitored before transmitting, it is necessary to use a line switched function available on standard remote control units. This reduces the number of functions possible for other purposes. In the case of DC switched remote controllers, this normally prevents the use of all other outgoing functions (talkthrough switching etc.) except send/receive.

By using tone switched functions, additional facilities can be made available, enabling send/receive, tone squelch inhibit and, say, talkthrough to be included.

4.3   Remotely Controlled Systems over Radio Links. With this mode of control, the same general method of inter-connection to the transmitter and receiver units applies, thus ensuring that whenever the transmitter radiates, the tone automatically modulates the carrier. Similarly, the receiver audio output to the link transmitter is only applied when a matching tone modulated carrier is received. The arrangement is shown in Fig. A17.1.

Unfortunately, with this mode of control, the receiver tone squelch can be inhibited only by tone methods. Using controllers with one or two facility tones is a possible method. In a multi-station control complex, additional tones must be employed.

Where the control is over multiplex links, the additional tones controlling the inhibit function must be placed in the upper section of the band 300–3400 Hz. The speech band must be limited to say, 2500 Hz by filters to avoid overhearing the tones and to prevent speech operation of the tone circuits.

5.   Use of Tone Lock as in "All Call" Selective System

In this system, shown in Fig. A17.2, tone lock is included to reduce the amount of "chatter" which is heard on all mobiles during operation of a normal system.

If there are, say, 30 mobiles in use and messages are spread fairly evenly throughout the system, for a considerable period of any hour each mobile is subjected to unwanted messages. By reducing this "chatter" to calls and wanted messages, the overall effect is less tiring for all concerned.

From Fig. A17.2, it can be seen that, assuming no sub-audio tone is being received by the mobile, and the over-ride switch is open, the loudspeaker is cut off and no messages can be heard. When the handset is removed from the rest or when the over-ride switch is operated, the reverse occurs and all messages are heard.

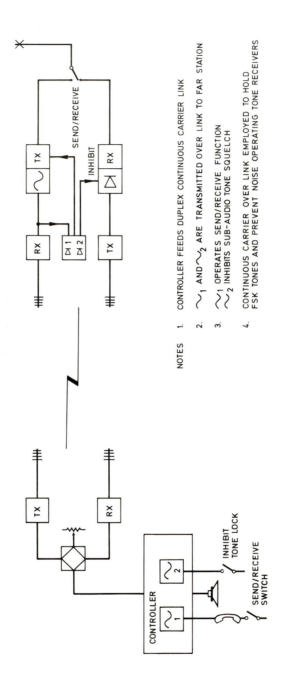

Fig. A17.1 Tone Lock over Radio Link

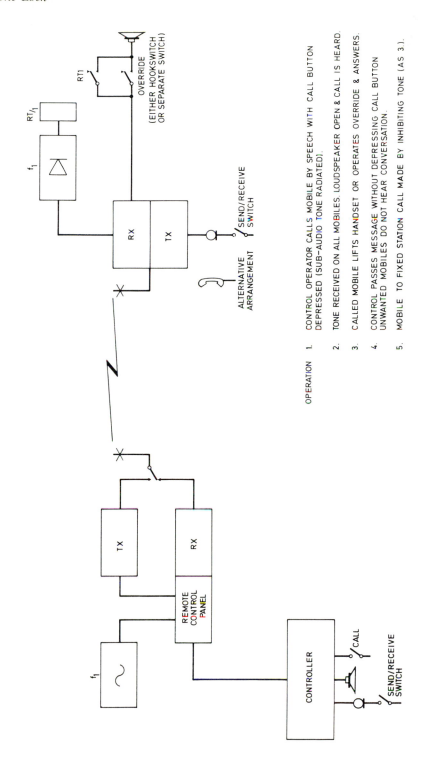

OPERATION

1. CONTROL OPERATOR CALLS MOBILE BY SPEECH WITH CALL BUTTON DEPRESSED (SUB-AUDIO TONE RADIATED).

2. TONE RECEIVED ON ALL MOBILES. LOUDSPEAKER OPEN & CALL IS HEARD.

3. CALLED MOBILE LIFTS HANDSET OR OPERATES OVERRIDE & ANSWERS.

4. CONTROL PASSES MESSAGE WITHOUT DEPRESSING CALL BUTTON UNWANTED MOBILES DO NOT HEAR CONVERSATION.

5. MOBILE TO FIXED STATION CALL MADE BY INHIBITING TONE (AS 3.).

Fig. A17.2 Use of Tone Lock for Calling Purposes

By leaving the mobile in the first condition (loudspeaker muted) the user is relieved of all message chatter, and only receives actual calls when the control operator switches the sub-audio tone into circuit.

Any speech heard from the loudspeaker is a call to a mobile. By lifting the handset or operating the over-ride switch the wanted mobile can reply. All other mobiles ignore the call and the subsequent message "chatter" is not heard by them.

Thus, any speech on the channel must be a call, and should be observed.

The system can be over-ridden continuously in any mobile, if so desired.

6. Selective Operation of Talkthrough

The talkthrough facility is a popular and useful method to enable mobile-to-mobile operation, via a favourably sited repeater, available to users employing two frequency operation. It is shown in Fig. A17.3.

Two basic versions are possible. The first is the controlled type where an operator has means of switching off the facility. The second is known as "free running" i.e. the facility is always available, and cannot be switched off by an operator.

Thus, in the second case, no control can be exercised over the facility at any time. The controlled version also suffers from this disadvantage if switched to talk-through and left unattended, e.g. at night or at any other time when an operator is not available. During such periods, any user of the channel within range triggers the repeater, and often, under conditions of anomalous propagation, the repeater can be operated by distant mobiles.

This unavoidable operation can cause considerable interference over large areas and P & T organisations in many countries are restricting the use of such devices.

One method of reducing unintentional operation of other networks is to include a sub-audio tone decoder in the triggering circuit. The frequency of this unit is peculiar to the user, and all mobiles and fixed stations of the network are equipped with sub-audio encoders. Other networks are not fitted with the same tone frequency, and, therefore, the talkthrough station cannot be triggered by them.

Further development of this method can include different tone frequencies for different repeaters in a given system. By having compatible tone units in each mobile and fixed station associated with the repeater, each can select a particular repeater for optimum coverage.

7. Conclusions

The use of tone lock and the control of the functions needed must be considered carefully. Obviously, where possible, a locally controlled transmitter and receiver combination will simplify the solution.

Where remote control facilities are needed, the simple system requiring send/receive and tone lock inhibit functions can be achieved by either DC or tone methods using standard practice. If, on the other hand, a second function is required, the use of tone switching presents the easiest answer.

Ultimately, tone switching may become mandatory in many countries with a result that the DC switching technique may become redundant. This should be borne in mind when initially evaluating the system.

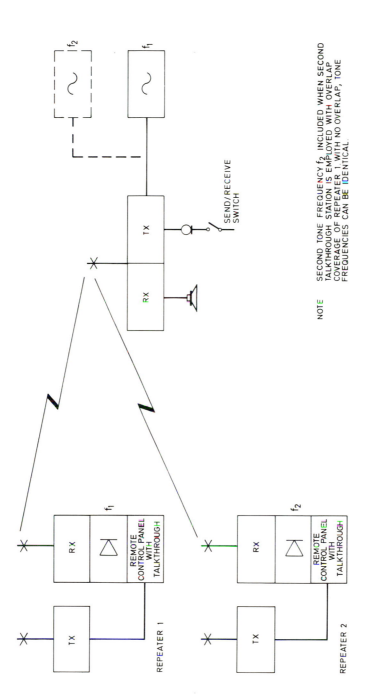

Fig. A17.3 Tone Operation of Talkthrough Repeater

Annexe A.   Remote Control Switching

1.   Remote Operation by Tone Switching

The existing practice of incorporating DC controlled remote operation in systems is becoming less favoured by many P & T organisations. Many are now in fact specifying VF tone operation of radio telephone systems over line pairs, and it is considered probable that the practice may become widespread in just a few years.

Consequently, it is considered that the use of tone switching, far from appearing as a "special" method, is rather one which should be encouraged.

Obviously, correct tone levels, impedance matching, etc. are of major importance in minimising inter-circuit crosstalk, spurious operation and other ill effects. These problems could be minimal compared with those arising from line unbalance, clicks and other noises caused by DC switching or distortion at the low frequency end of the audio spectrum caused by line loading, etc.

One major advantage of using tone switching is that DC repeaters and associated techniques are not required when passing through line sections employing amplification.

Another advantage is that, with tone switching, line reversals do not introduce the problems associated with line to earth DC switching.

Over radio links, either single channel or multiplex, tone signalling is of course essential and this is yet another reason for standardisation in control techniques.

2.   Remote Operation by DC Switching

In spite of the disadvantages outlined, DC switching will probably still be employed in many parts of the world for a considerable time. Its main use lies in circuits under the control of the user, whilst, for initial cost economy, it shows considerable advantages over the more complicated tone systems.

There is, of course, no reason why the two systems cannot be combined and obviously, where DC conditions already exist, i.e. in circuits associated with the telephone network (both manual and automatic), the only means by which additional switching can be incorporated is by adding tone circuits.

Appendix 18.   Radio Propagation – The Problems of Fading Over Radio Links

1.   General

Radio circuit outages usually fall into one or more of three main categories:

(a)   Failure of equipment owing to internal faults caused by design short-comings, component failure, etc.

(b)   Failure of one or more elements of a system owing to unforeseen events. This includes failures caused by lightning, flood and sabotage.

(c)   Atmospheric changes causing modifications to the signal path. At worst, these changes can cause the system to cease working. The usual effect is an overall deterioration of the signal-to-noise ratio for periods with depths determined by the fading mode.

The first problem can obviously be minimised by equipment design improvements, adequate component ratings, and the use of stand-by equipment. The causes of failure in the second category can be eliminated by measures such as suitable siting, lightning protection and security.

Although caused by atmospheric changes, the last problem can be alleviated by suitable system design. The following paragraphs are intended as a guide to improvement in this respect.

2.   Idealised Radio Path

To achieve a substantially constant path attenuation, certain stable conditions are required. They include an ideal constant climate with sufficient turbulence to minimise multi-path propagation, suitable siting of the terminals with no path obstructions and adequate Fresnel clearance, intervening terrain which is sufficiently broken to eliminate ground reflections and a path short enough to ensure that any rain or snow does not cause additional path losses at the frequency employed. Such a situation is illustrated in Fig. A18.1.

To complete the ideal case, external reflecting bodies such as aircraft should not be permitted within the area where they are likely to affect the received signal.

Fig. A18.1 Idealised Radio Path

Obviously such conditions seldom exist and therefore the system must include sufficient margin to cope with the foreseeable variations based either on calculation or on measurement over the paths concerned.

3. Causes of Fading

Four basic fading modes can be isolated. All are considerably modified by the choice of frequencies with a general increase in fading as the frequency rises.

3.1 Changes in refraction. Refraction is governed by the rate of change of the dielectric constant of the atmosphere with height. It determines the degree of bending of the radio beam. Bending of the radio beam can alternatively be considered equivalent to a straight radio beam travelling over an earth's surface having a modified curvature.

The average rate of change of dielectric constant (approx. $2 \cdot 4 \times 10^{-8}/$ ft) modifies the earth's radius to a figure of $K = 4/3$ and this is considered the normal figure for most calculations, particularly below 500 MHz. (When $K = 1$, true earth radius applies.)

If the rate of change of the dielectric constant increases to a value of approx. $1 \times 10^{-7}/$ ft, the earth can be considered electrically flat and K equals infinity. At other values of dielectric constant change, the bending can become such that the beam will tend to leave the earth's surface and, over certain lengths of path, this may result in a considerable or even a complete loss of signal. The upward bending can also result in the return of the signal by reflection from the upper atmosphere and further reflections can occur alternately by reflections from the earth and atmosphere.

This effect, known as ducting or anomalous propagation can often cause a high level of long distance interference. Owing, however, to the difficulty of predicting anomalous propagation, the method cannot be used as a reliable means of providing long distance communication.

Fig. A18.2 shows the influence of different rates of change of dielectric constant. The propagation paths are as follows:

(*a*) Normal path with average rate of change of dielectric constant (Generally $K = 4/3$).
(*b*) Path with reduced rate of change of dielectric constant.
(*c*) Path with high rate of change of dielectric constant giving long range propagation.
(*d*) Path with very high rate of change of dielectric constant exhibiting "bouncing" between upper atmosphere and earth, also often results in lost signal at receiving terminal.

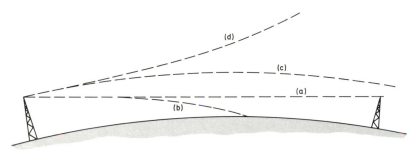

Fig. A18.2 Changes in Refraction

(*a*) is the normal path, (*b*) and (*d*) result in a weaker or lost signal; (*c*) gives a potentially enhanced signal, especially on long paths. Both (*c*) and (*d*) represent conditions in which long range interference is probable.

A reduction in the dielectric constant rate of change (below the average value where $K = 4/3$) will, of course, result in an increase in the earth's effective radius. Over a given path, the effect can cause a reduction or even a loss of signal due to the intrusion of obstacles or the earth's mass into the radio path. This is shown in Fig. A18.3.

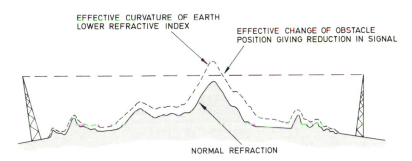

Fig. A18.3 Changes in Refraction with Reference to Obstacles

Therefore to minimise the fading range and effects due to changes in refraction, the following must be observed:

(*a*)   The path must have adequate clearance over the earth's mass or likely intervening obstacles. The clearances must be observed over the range of values of K likely to be experienced in the area concerned.

(*b*)   If upward bending is a known phenomena in the area, then the hops must not be extended too far.

(*c*)   Where anomalous propagation is likely, long distance interference must be minimised by suitable frequency allocation.

(*d*)   In the areas of the world where fading is likely to be caused by changes in the value of K, the periods of signal deterioration or loss are often fairly long – up to several hours. It is therefore essential, over such paths to include sufficient margin in the reliability calculations to enable a circuit to be achieved during these periods.

3.2   Multipath Propagation. When a transmitted signal travels by more than one path to the receiving terminal, the possibility of the signal over one path arriving in phase with signals over the other paths, and remaining so, is extremely remote.

Consequently the resultant signal may vary from enhancement of the normal level to a partial or even complete cancellation.

Fig. A18.4 shows the mechanism of multipath propagation.

It can be seen that deeper fades can be produced by propagation over two paths than when signals arrive over three or more paths. This is because the chance of complete cancellation is much less over the larger number of paths. Multipath propagation in the atmosphere is generally produced by sudden changes in the slope of the dielectric constant relative to height and such points of transition are often caused by changes of terrain as from land to water or from open country to city. Paths parallel and adjacent to a coastline can also be affected quite markedly by multipath conditions.

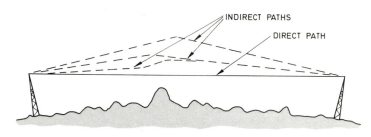

Fig. A18.4 Multi-path Propagation

Quiet, windless and foggy nights will often produce these conditions owing to lack of turbulence.

Fading caused by multipath propagation is likely to be frequent and short and so has a considerable effect on links carrying high speed data unless the system is specially designed to minimise the problem.

To ensure that an adequate signal arrives at the receiver involves a method virtually equivalent to an artificial form of multipath propagation. It is obviously impossible to guarantee that sufficient paths will exist at all times through the atmosphere to provide a normal signal with minimal deep fades. However, the chance of cancellation can be reduced by transmitting one or more additional parallel paths, using suitably related frequencies calculated to avoid trough coincidence. This method, known as frequency diversity, relies on the short duration of multipath fading and the natural phase differences possible with suitably related frequencies. By selecting frequencies based on the ratio of sequential integers, the possibility by coincidence is reduced considerably.

As an example, frequencies in the ratio of 8 to 9 may be used, such as 80 and 90 MHz, 400 and 450 MHz or 800 and 900 MHz.

Reference to Fig. A18.5 illustrates the possible advantage to be gained in reduced fading depth as a function of frequency separation.

3.3   Fading due to reflection changes. This form of fading often occurs over short paths when the intervening terrain has a reasonably high coefficient of reflection.

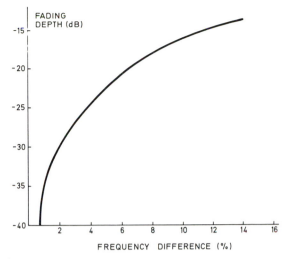

Fig. A18.5 Maximum Fading Depth with Frequency Diversity System assuming Two-path Propagation

Paths over water or flat marshy areas are typical examples, whilst even flat desert areas can cause quite high levels of reflected signal.

The received signal is produced by the addition of the direct and reflected components. When the received signals are in phase, the resulting signal may be twice that of the direct path alone. The resulting signal is reduced at other phase differences until, at the point where the signals are opposite in phase, minimum signal is available.

The effect is obviously at its worst when the direct and reflected signals are equal in amplitude.

Fading on an otherwise suitable path is usually caused by the following:

(*a*)   Over a sea path, tidal changes can cause phase changes between the direct and reflected path as shown in Fig. A18.6.

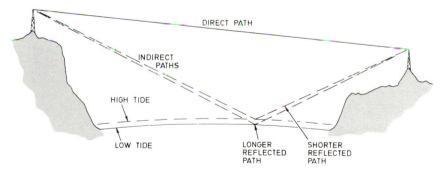

Fig. A18.6 Reflection Changes – Tide Effect

(*b*)   Reflection changes over the path caused by changes in refraction. Where the result of fading caused by reflection changes is likely to be serious, various methods of minimising the effect can be implemented:

(i)   Siting the terminals so as to introduce obstacles in the reflected path whilst maintaining a clear path for the direct beam. With water reflection, siting back from the edge using a land mass as the obstacle to reflection is often feasible. Fig. A18.7 shows this scheme.

(ii)　*Spaced diversity techniques*

The general layout is shown in Fig. A18.8. Two receivers are used with the two antenna systems spaced in the vertical plane by what is known as half lobe spacing as shown in Fig. A18.9. This ensures that one receiver can receive both paths in phase when the second receiver

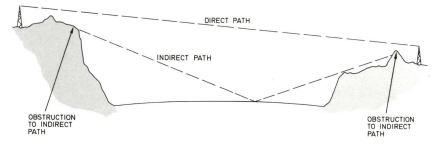

Fig. A18.7 Reflection Changes – Minimising Tide Effect by Siting

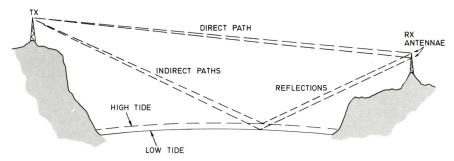

Fig. A18.8 Reflection Changes – Minimising the Effect by Space Diversity (One Direction only)

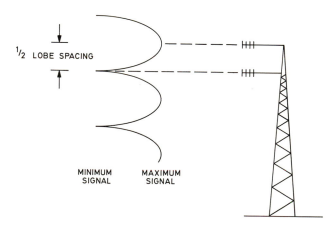

Fig. A18.9 Reflection Changes – Space Diversity Technique

is fed with out of phase signals. As conditions change, the inputs to the receivers change but can never reach a state where both receivers are zero at the same time. For optimum conditions, the reflection point should be well below the direct paths: space diversity over grazing paths will not be as effective as over paths with adequate clearance.

3.4 Fading caused by additional short term paths losses. This form of fading is a function of frequency. It is usually produced by changes in the medium existing between the terminals and embraces such effects as heavy rain storms, snow and fog. Fig. A18.10 shows the problem.

At VHF, very little effect is usually experienced and it is not until the UHF and microwave spectrum is reached that the phenomenon starts to be troublesome.

It has been stated (Ref. 6) that, in moderate climates, rain attenuation is of the order of 2–4 dB at 2 GHz and increases to 10–15 dB at 6 GHz. At higher frequencies, tropical rainstorms can often cause a complete loss of signal.

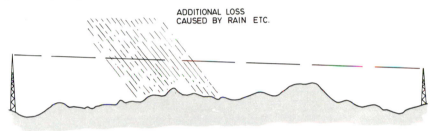

Fig. A18.10 The Effect of Rain, Snow, etc.

Fog and cloud in the radio path often causes some fading at the higher frequencies and can occasionally increase the signal level at the lower frequencies by causing beneficial refraction changes. Only frequencies above 10 GHz are likely to suffer from deep fades due to increased losses.

The main remedy for this effect is to use either lower frequencies or higher power.

3.5 Long term reduction or loss of signal. Whilst this form of loss can hardly be termed a fading effect it is as well to mention the problem at this point.

Where the terrain associated with a given path exhibits seasonal physical changes such as heavy summer foliage in forested areas and where the signal path is likely to pass directly over or through such an area (Fig. A18.11), then adequate tolerance in the calculation or measurement is essential.

Fig. A18.11 The Effect of Trees, etc., in or near the Path

A midpath obstruction is likely to cause seasonal variations owing to foliage growth. A similar obstruction closer to either terminal will have a similar effect but of greater magnitude. In this case, the losses will be similar to those shown in Fig. A18.12, given by Bullington.

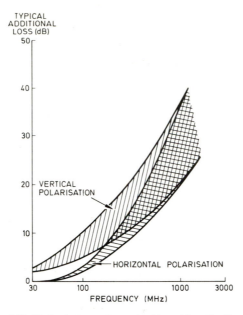

Fig. A18.12 Path Loss through Absorbing Bodies – Trees

Where the trees are below the direct path, some modification may be necessary to the calculations owing to reflection changes.

Medium to long term loss is often caused by buildings in the path particularly if they are constructed of a material which absorbs moisture during rainy seasons. Fig. A18.13 shows the likely variation of attenuation under wet and dry conditions.

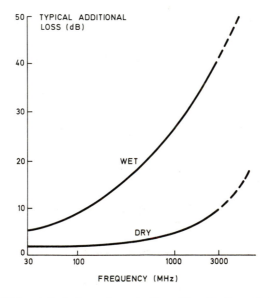

Fig. A18.13 Additional Path Loss through Absorbing Bodies – Brick or Stone Wall

Further care must be taken when planning systems using point-to-point links in higher frequency bands. The construction of buildings in or near the path can often cause a complete loss or, at the best, a considerable reduction in signal level and if the construction takes place some time after the installation of the radio equipment, considerable confusion can arise.

To eliminate or minimise the effect, one must increase the tower height, or relocate the site to avoid the obstruction.

4.   Conclusions

Fading is a phenomenon which cannot unfortunately be eliminated, only minimised. The foregoing notes suggest methods of reducing the range of fading likely to be experienced over any link path.

The degree of fading can, to some extent, be calculated. Alternatively a fixed margin can be included in the system calculations. In the latter case, the margin should be estimated on existing evidence or preferably on the basis of a prolonged survey using parameters similar to the final system. Such a survey cannot, in many parts of the world, produce adequate evidence of the type, depth and frequency of fading unless covering at least 24 hours throughout all seasons of the year. Such a survey is normally not possible and therefore it is often inevitable that fading margins must be based on information derived from previous similar links together with such adjustments dictated by the climate and terrain.

Appendix 19. Calculation of Receiver Protection Requirements

This comparatively simple task can be further helped by a suitable typical set of curves showing the approximate rate of improvement as the tuned frequency differs from that of the interfering signal. To cater for all likely types of system – power output, antenna separation, etc. Fig. A19.1 has been compiled. It applies in general to the VHF band only, and is approximate only. The attenuation is shown in dBW (i.e. dB relative to 1 watt), so adjustment is needed to cater for actual output power (e.g. an increase of 14 dB for a 25W transmitter).

Its use is best shown by typical examples:

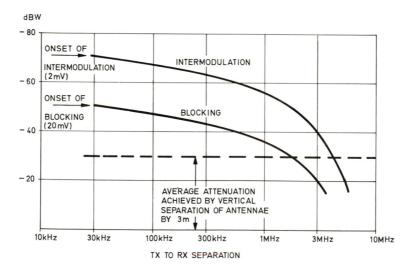

Fig. A19.1 Typical Receiver Protection Requirements

(a) Let us consider the general system requirements if a 10 watt 80 MHz transmitter is co-located with a receiver and that the greatest possible Tx to Rx antenna vertical separation is 10 feet (3 m). Fig. A19.2 shows the isolation obtainable with this separation as 30 dB. Only a single Tx Rx pair is involved, so that blocking is the limiting factor; intermodulation effects obviously do not occur in a single unit case.

Examining Fig. A19.1 and following the –30 dB line across the chart until it meets the blocking curve, it can be seen that this point is reached at a Tx to Tx spacing of 1·8 MHz. However, as the Tx output power is 10 watts, either the total curve must be raised bodily by the equivalent of 10 dB or the easier solution of deducting 10 dB from the antenna isolation can be applied.

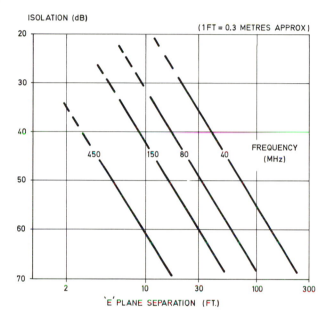

Fig. A19.2 Antenna Isolation as a Function of Vertical Separation

If the latter is adopted, the intercept point becomes 3 MHz and this is the minimum Tx to Rx spacing to achieve the desired isolation under the conditions given.

(b) Next, let us consider that only 50 kHz exists between a transmitter and receiver at 80 MHz. Again 10 watts (+10 dB relative to 1 watt) output power is considered with single Tx to Rx units only.

The curves show that, at 50 kHz, only 3 dB isolation can be expected from the front-end selectivity of the receiver and therefore, below the curve, 49 dB is found to be required plus the additional 10 dB to account for the 10 watts of output power. Thus we need 59 dB of space attenuation which from Fig. A19.2 shows that some 60 feet of vertical separation is required (a difficult requirement to meet on the average mast), or, alternatively but more likely, the transmitter and receiver must be separated by 1500 feet (approx. 480 m). Fig. A19.3 shows this parameter relative to path attenuation.

(*c*)   Assume in this case communal siting arrangements with 25 watt transmitters, 30 dB average attenuation between antennae and a frequency plan by which overloading of the receivers and certain on-channel intermodulation products are likely to be troublesome. Let us consider the 150 MHz band in this case.

First of all let us deduct the 14 dB introduced by the use of 25 watt transmitters from the 30 dB antenna isolation. This can now be shown to be re-adjusted to 16 dB. The intercept point of this figure with the inter-modulation curve – in this case we must use the I.M. curve instead of the blocking curve as more than one Tx Rx combination is involved – shows a minimum frequency spacing of 6 MHz is required between any Tx and any Rx likely to produce intermodulation products.

(*d*)   If we now take the parameters of (*c*) above but substitute 25 kHz channel separation as, for instance in the International Marine Plan, we show only a 1 dB front-end protection, leaving 71 dB plus the 14 dB power gain i.e. 85 dB.

Horizontal separation between transmitter and receiver sites as shown by extrapolating Fig. A19.3 indicates that 8000 to 9000 feet is necessary for adequate isolation (approximately 2600 m). In practice, this figure tends to be high as the path attenuation at distances over a mile will increase at a

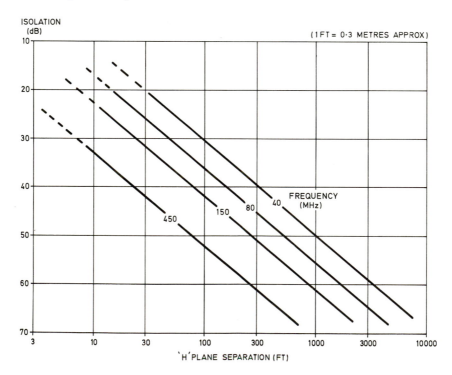

Fig. A19.3 Antenna Isolation as a Function of Horizontal Separation

higher rate than 6 dB per doubling of distance and thus, from experience, the more likely order of separation would be around 5000 feet (1500 m).

The curves are intended as approximations only and indicate the order of protection necessary or likely to be achieved. Obviously, more accurate calculations can be made but are outside the scope of this Appendix.

Appendix 20. High Gain Omni-directional UHF Base Station Antennae – The effect on channel re-use and signal levels in the immediate vicinity

From time to time one is asked about the performance of UHF base station antennae with gain in excess of the 5·5 dB. The following notes are intended to assist those who may consider their use. There are two main designs of omni-directional antennae, collinear and dipole arrays. Usually, collinears are narrow band but can be mounted in slim fibreglass tubes; dipole arrays are broader band but are less aesthetically pleasing.

1. Beam Width

The vertical beam width of an 8 to 10 dB omni-directional antenna is about ±4° to the 3 dB (half power) points. The shape of the polar diagram is such that, beyond 4°, the gain falls very rapidly and at ±5 to 7° one can expect a complete null. If the antenna is mounted 200 feet above the surrounding terrain, the shadow area not covered by the main polar diagram is about 1000 yards in diameter. Fig. A20.1 shows a typical case.

One of the problems in designing a high gain array is to maintain the polar diagram horizontal over the frequency range. A beam may tilt 2° over a 10 MHz range. Provided this tilt is negative i.e. towards the ground, this is not unsatisfactory, but if the tilt is positive, it can extend the shadow area to 2000 yards radius. It is thus obvious that serious lack of reciprocity between "go" and "return" can exist on a two frequency system, particularly when the degree of tilt changes rapidly with the relative small change in frequency associated with such systems.

The foregoing facts pre-suppose that the antenna is installed absolutely vertically. In practice this is extremely difficult to judge at the top of a high tapered tower. An installation tilt of 4° can result in no coverage of the main beam away

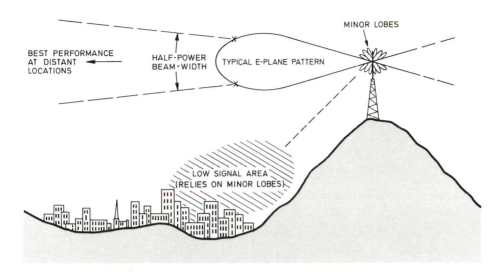

Fig. A20.1 Effect of Using High-Gain Narrow E-Plane Antennae on High Sites

from the tilt. One often forgotten aspect is that high gain antennae are long and therefore flexible. In strong winds the antenna or the structure on which it is mounted may flex sufficiently to tilt the polar diagram.

2.   Minor Lobes

Any high gain antenna will exhibit a bunch of narrow lobes in directions away from the major lobe. Communication close to the antenna site relies on these minor lobes either directly or from reflection. Although the antenna designer normally tries to design out these minor lobes, since they represent gain lost from the main lobe, they are often anomalously essential for the system operation. Minor lobes are largely unpredictable and their amplitude and beam width can vary considerably over the frequency range of a particular antenna. They are, therefore, unreliable.

3.   Manufacturers' Claims

There is no fixed convention for quoting antenna parameters. Some manufacturers quote the maximum gain in the frequency band in bold type. Examination of the small print sometimes reveals lower gain at the band edges. Other manufacturers quote the minimum gain over the frequency range; the inference here is that the gain is almost certainly greater somewhere near the centre of the band. Only a few manufacturers come to terms with beam tilt by showing curves of tilt against frequency. The practice of using an isotropic reference gain may still be used and any gain figures need to be carefully studied. Vertical beam width can be used to calculate the gain since there is a mathematical correlation. If two manufacturers claim the same gain but different beam widths, the one with the narrower beam width is probably nearer the truth.

All manufacturers now appear to quote gain with respect to a halfwave dipole which is meaningful in the radio telephone field.

4. Conclusions

8 to 10 dB gain omni-directional antennae can give a worthwhile extra margin to a scheme but, unless proper selection and installation is carried out, the results can be catastrophic, both from the loss of signal in the immediate vicinity and the tendency to enhance signal levels at distances at which the possibility of channel re-use must begin to be considered.

Appendix 21.   Typical Mobile Radio Users

Posts & Telegraph Administrations

Telephone Line Maintenance
Public Telephone Maintenance
Radio Bearer Link Maintenance
Mail Security Services
Public Radio Telephone Services
Rural and Island Low Capacity Radio Links
Post Buses, Parcel Delivery and Collection Services

Ministry of Transport

Railways
Road Haulage Services
National Bus Services
Harbours and Docks
Inland Waterways and Canals
Airport Services
Motorway Control Services
Underground Services
Cable Cars

Government Departments
Police
Civil Defence Organisations
Coast Guards
Customs and Excise
Lightships and Lighthouses
Lifeboat Services
Civil Aviation Authority
Airport Maintenance and Control Services
Ministry of Works and Public Buildings
Food & Drug Administration
Ministry of Health Ambulance Services

Ministry of Agriculture, Fisheries and Food
Farm & Veterinary Inspection
Fishing Patrols
Blood Transfusion Services
National Laboratories
Forestry Conservation and National Parks
Ordnance Survey and Cartographic Services
Meteorological Offices and Observatories
Treasury and National Mint, Gold Storage, etc.
Ministry of Justice – Courts
Prisons
Taxation Investigation

Broadcasting Authorities

Studio and O.B. Cueing and Control Communications
Transport Communications
Maintenance Service Personnel

Local Government

Detention and Correction Centres
Fire Service
Salvage Corps
Civil Defence
Highways, Bridges and Tunnel Maintenance
Parks and Playing Fields Maintenance
Educational, Transport and School Buses
Beach, Mountain and Cliff Rescue Services
Traffic Wardens
Probation Officer Services
Street Lighting and Traffic Signs Maintenance
Sewage and Drainage
Refuse Collection and Street Cleaning
Public Health Inspection
Flood and Snow Control Services
Infestation and Pest Control

Utilities

Electrical Generation and Distribution Services
Atomic Energy Establishments
Gas Production and Distribution Services
Coal Mining and Distribution
Water Catchment and Distribution
Hydro-electric Water Supplies
Irrigation, River Conservancy and Control

Transport

Road Haulage and Trucking

Bus Companies
Hovercraft
Haulage Security Services
Parcel Collection and Delivery Services
Taxi Cabs
Private Hire Cars
Funeral Services
Emergency and Breakdown Services
Ferry Services, Inland and River
National Motoring Clubs (e.g. A.A. R.A.C.)

Service Industries

Home Appliances (Refrigerators, Washers, Cleaners, Mixers, etc.)
Plumbers
Electrical Repairs (Lifts, Escalators, Conveyors)
Typewriters and Office Equipment
Computers, Comptometers, etc., Data Storage
Industrial Machinery and Machine Tools
Radio, TV, Audio Equipment
Private Telephone Repair Services
General Household Repairs
Painting and Decorating
Laundry and Dry Cleaning
Burglar and Fire Alarm
Extinguisher Services and Repair
Traffic Light and Road Sign Services and Repair
Automatic Vending and Gaming Machine Maintenance & Collection
Music Equipment Maintenance
Oxygen and Industrial Gas Distribution
Air Conditioning Equipment
Public Address Equipment

Industrial

Works and Factories
Quarries and Mines
Oil Refineries, Wells and Tank Farms
Chemical Works
Factory Security Guards
Works Fire Service
Works Medical Services
Civil Engineering (Works and Buildings)
Highways, Bridges, Tunnels, Dam Construction
Ready-Mix Concrete Services
Sand, Gravel and Aggregates Extraction and Delivery
Brick, Stone and Prefabricated Structure Services
Machinery Services (Plant Hire and Servicing)
Surveying

Agricultural and Food Industry

Farm and Horticulture
Agriculture Machinery Installation and Maintenance
Crop Spraying Services
Artificial Insemination Services
Slaughtering and Collection
Crop and Produce Collection
Estate Management
Game Keepers and Water Bailiffs
Agricultural Plant Hire
Hatchery Service
Food Processing and Mills
Fish Docks and Processing

Business and Management

Executives and Managers
Marketing and Sales Staff
Market Survey Staff
Insurance Agents
Estate Agents
Banking and Savings Staffs
Private Detective and Enquiry Agents
Cash, Bullion and Precious Stone and Metal Security
Office, Works and Site Security Staffs
Accountants and Auditors

News, Information and Entertainment

Newspapers and Reporters
Films – Location and Studio Communications
Films – News Services
Press Photographers
Newspaper Delivery Services
Stage and Theatre Communication Services
Sports Arenas
Car Parking Control
Horse Racing and Course Control
Motor Cycle Rally and Race Control
Outdoor Sporting and Athletic Events
Car Racing Circuits

Aeronautical – Ground Based

Private and Municipal Airport Services
Control Air Transport Operators
Airline Service and Maintenance
Airport Apron Services
Gliders
Helicopters

Private and Executive Flying
Flying Schools and Clubs
Airport Police and Security
Airport Crash, Fire and Medical Service
Crop Spraying

Public Radio Services

Public Radio Telephone
Public Radio Paging Services
Public Message Bureau Services
Co-operatives of Users with Common Carrier Handling Facilities
Community Radio Services
Emergency Medical Treatment Services

Forest Products

Logging
Mills
Transport & Storage
Forest Rangers & Fire Watching Staff

Marine (Land Based)

Harbour
Warehousing and Servicing
Dock Police and Fire Services
Lighters and Barges

Medical Services

Ambulance Services
Doctors
Specialists
Medical Auxiliaries
Hospitals and Clinics
Emergency Drug, Blood and Medical Gas Distribution
Transport and Allied Services
Midwives and District Nurses
District Nursing and Midwifery Services

International Organisations

United Nations Organisation and its specialist Services
World Meteorological Organisation (W.M.O.)
World Health Organisation (W.H.O.)
International Civil Aviation Organisation (I.C.A.O.)
International Aid Foundations
Medical Missions
Relief Organisations
United Nations Organisation (U.N.O.) – Peace Keeping Forces

General Public

Commuters – on foot or vehicles
Private Transport
Private and Social
Sport and Recreational
Security and Breakdown Services

Appendix 22.   The Location of Mobile Radio Systems on Sites Occupied by High Power Transmitters of Unrelated Users

1.   General

For systems designed for wide area coverage, the normal arrangement is to elevate the antennae by choosing a high site positioned so as to cover the required area. As postulated elsewhere, this method is subject to some reservations, particularly because the re-use distance of such a site is much greater than if the area were covered by a number of lower sites. However, it is often the most economic system, indeed maybe the only system, which can be utilised in some cases.

Unfortunately this solution is also often adopted by other services; in particular, for example, the radio and TV transmission authorities. So one must often consider co-siting the base stations of land mobile radio systems with high power radio or TV transmitters to obtain the desired coverage.

Clearly, then, in these circumstances, the mobile system radio receivers on the site are liable to be upset by the high power transmitter(s).

2.   Characteristics of a typical high powered TV transmitter

A 625 line TV transmitter may have not only an extremely high peak power level (as much as 1000 kw) but also a total bandwidth of between 6 and 8 MHz. Thus, at close range, the effect upon mobile system radio receivers at frequencies quite widely separated from such a transmitter can be quite marked.

Let us convert the figure of 1000 kw into a voltage across 50 ohms (the normally recognised input impedance of a mobile radio receiver). This is nearly 7100 volts. When considered against an *effective on-channel* unwanted voltage of 20 mV which a normal mobile radio receiver can accept without undue degradation of the wanted signal, the unwanted signal needs to be attenuated by 110 dB. Furthermore, bearing in mind that the TV transmitter occupies about 6 MHz of bandwidth, with a not particularly sharp modulation response cut off slope, then clearly, for reasonable isolation, co-sited mobile radio and TV systems must be separated by tens of MHz. Alternatively, the space attenuation, again assuming that the nominal TV modulated band is outside the normal mobile radio RF passband, must tend towards 110 dB, equivalent to tens of miles between high sites using nearby frequencies.

Fortunately, all TV transmitters do not radiate powers as high as 1000 kw, but even, say, 5 kw the voltage across 50 ohms is 500 volts, a reduction of only 23 dB on the 1000 kw level (from 110 to 87 dB).

3. Sharing of Sites

Evidently, the sharing of a site by TV and mobile radio systems is a dangerous proposition without considerable frequency spacing between the systems.

For instance, low or even high band mobile radio systems may operate perfectly satisfactorily on sites occupied by UHF TV transmitters, although the reverse may not necessarily be true. In the latter case, TV transmitters in the 41·5 to 69 MHz or in the 174 to 216 MHz bands could provide quite high levels of spurious harmonic signals in the 400 to 500 MHz band used by mobile radio. Even if the TV harmonics are 80 to 90 dB down on the main carrier, the level is still sufficient to cause chronic interference.

4. Unwanted Products

Although constrained by international standards including signal modulation levels and carrier noise, TV and radio transmitters are likely to give rise to inter-modulation products in nearby receivers. These may be generated not only from individual carriers but also from sidebands.

Sideband noise will, in effect, add to carrier noise and produce a very wide spectral bandwidth which can cause intermodulation in the receivers unless the level is below about 2 mV.

VHF broadcasting *networks* are also sources of interference, since at least two and often three or four such stations are co-sited. Furthermore, they generally have equally spaced frequency differences giving a regular succession of products. Fig. A22.1 shows some of these products. The effective deviation of the inter-modulation products is a function of the total instantaneous deviations of the contributory signals. Since the deviations of harmonics will be multiples of the fundamental, the deviations of the corresponding products will be similarly augmented.

Such products are often of relatively low level from the transmitter complex (transmitted intermodulation). However, if mobile system receivers on nearby frequencies are close to the broadcasting site, some interference will be generated in those receivers. Again, being relatively broadband (200 kHz channel spacing and 75 kHz deviation), the spread of the broadcasting transmitters will be quite wide.

5. Precautions

Although the foregoing paints a rather gloomy picture, co-siting of mobile radio

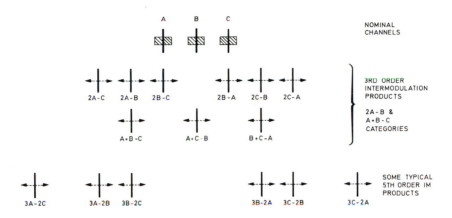

Fig. A22.1 Three-station Intermodulation Pattern from FM Broadcast Station (Equal Carrier Spacing)

systems and broadcasting systems has been undertaken in the past and with some success. Any proposed site must be analysed and the above points observed before risking the expense of installation and possible later re-siting. It is as well to perform a monitoring check on the site using the frequencies to be allocated and the antenna systems in their approximate final position.

Fortunately, any antenna system for a mobile radio system is unlikely to approach the height of the broadcasting antenna array. With the high gain of such an array, the E-plane pattern at heights some distance below the broadcasting antenna is likely to provide considerable attenuation relative to the main beam (see Fig. A22.2). Therefore some further assistance in reducing the unwanted signal level will be obtained.

Another problem is direct breakthrough into the receivers and for that matter into the mobile system fixed transmitting equipment. Unless adequately screened, direct breakthrough into certain stages of the receiver can completely nullify any filtering precautions included in the antenna system. RF breakthrough into the modulation circuits of the mobile radio system fixed transmitter can produce unwanted distortion and even unwanted modulation on the mobile system carrier. Often a screened room or cabinet assembly is required to minimise the effect; high grade co-axial cable with a solid outer screen should be used rather than mesh with, possibly, double-screened cable almost certain to be essential in some situations.

Whether the same antenna mast or tower is used for both systems depends upon the site policy. The use of the same tower has certain advantages; the fact that the mobile system antenna is well and truly in the null of the broadcasting antenna pattern is possibly one of the main gains.

However, the structure may well introduce a marked loss of mobile signal in the rearward direction and thus the use of a *separate structure* on the same general site may be necessary. Even so, a large television antenna tower may affect the coverage of the mobile system in the direction of the tower. This effect is described at length in Appendix 23. Additionally, of course, the attenuation loss between the mobile system antennae and broadcast antennae will be less and the mobile system may need further filtering aids adding to combat the difference.

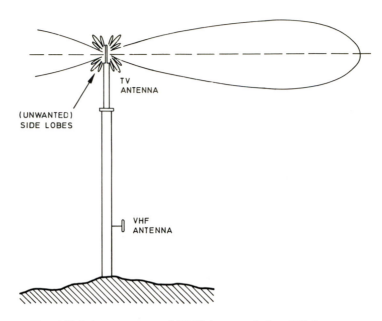

Fig. A22.2 Arrangement of VHF Antenna below TV Antenna

6. General Conclusions

Whilst TV and radio sites may provide all the essential requirements for the wide coverage needed from a particular mobile radio system, the arrangement must be approached with care.

All the frequencies in use on the site both before and after the possible installation of the mobile radio system must be analysed and signal levels calculated in relation to the frequencies used, the radiated bandwidths of the high power transmitters, the intermodulation and harmonic possibilities and their likely levels, as well as both the economics of the system (including the cost of additional filtering, screening, etc.). The layout and building requirements (and associated costs) must be thoroughly investigated.

Bearing in mind the broadband characteristics of the broadcasting systems, one must also weigh the implications of the above against the limited number of costly mobile radio systems likely to be possible on such a site and the limited frequency bands in which operation will be satisfactory even with the precautions included.

The main conclusion is that the mixing of such systems on a common site or within a short distance of one another should be avoided if at all possible, both from the economic viewpoint and the need for the enforcement of extreme precautions. There are some cases in which TV transmitters in the upper UHF bands may be sufficiently free from certain of the effects quoted and therefore satisfactory results may be possible. With TV transmitters in the low bands, the problems are likely to be acute in many circumstances.

Appendix 23.   The Effect of Metal Support Structures on the Performance of VHF and UHF Antennae

1.   General

The true characteristics of an antenna are only obtained when it is located in *free space* which, in this context, is space clear of the earth or other bodies so that radiation from the antenna is propagated by direct path only. The separation between the antenna and any structure must, therefore, be in excess of a pre-determined minimum distance – usually defined in wavelengths – unless, of course, the modifying effect of the structure is required to provide a particular radiation pattern. Obviously, the lower the frequency and the larger the structure, the greater the separation must be if the effect on the pattern is to be minimal. If, however, the antenna approaches the earth or structure the effect on the characteristics, in particular on the radiation pattern, becomes more and more pronounced; whilst ultimately, at close spacing, changes to the antenna impedance will also become noticeable with obvious effect on the standing wave performance.

The following paragraphs outline the effect of various types of structure upon normal antennae and indicate means of minimising these shortcomings or, alternatively, of using them to modify the antenna radiation pattern to suit a particular application.

2.   Basic Patterns Obtained in Free Space

2.1   Half-wavelength Dipole (transmitting and receiving). This is the simplest form of antenna currently in use at VHF and UHF. The radiation pattern in free space measured in the *far field* takes the form of a doughnut (Fig. A23.1). The *far field* of an antenna is generally considered as beyond the distance $2d^2/\lambda$, where d is the antenna length and λ the wavelength, measured in the same units.

The half-wave dipole is normally used as a reference for gain in VHF and UHF

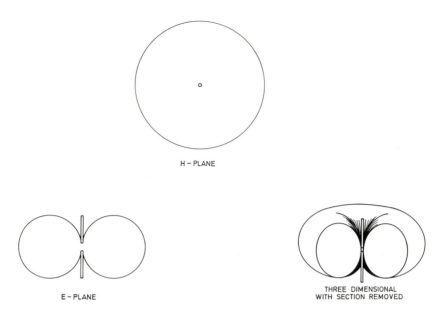

H − PLANE

E − PLANE

THREE DIMENSIONAL
WITH SECTION REMOVED

Fig. A23.1 Radiation Pattern of Dipole Antenna

antennae. The dipole has an effective gain of approximately 2·2 dB over an istropic radiator, which is a theoretical source of radiation having a spherical pattern with equal gain in all directions. The isotropic radiator is normally used as a standard reference at microwave frequencies.

2.2 Gain Antennae. This type of antenna uses phased driven elements and/or passive reflecting elements arranged to concentrate the radiation pattern into one or more required directions.

The total combinations possible are too numerous to outline in this Appendix. It can be seen however, from the examples in Figs. A23.2 and A23.3, that the effect of concentrating the pattern in a particular direction is to produce an effective gain in that direction. This increase in gain cannot be achieved without a reduction in gain in other directions and it is this effect which improves the performance of a directional antenna by limiting unwanted radiation in other direction(s). (Note that the diagrams have been normalised to show identical sizes of the main lobe.)

The best analogy illustrating the overall principle of obtaining gain by modification to the radiation pattern is that of a spherical balloon filled with a non-compressible liquid such as water. In its normal state it shows the pattern of an isotropic antenna with equal radiation in all directions. By compressing the balloon at opposite ends of a diameter, the shape becomes that of a doughnut and the diameter at right angles to the line of compression extends to give the effective dipole gain of 2·2 dB over an isotropic source.

By applying pressure at other points, the shape of the balloon can be altered to produce radiation patterns similar to those of directional antennae. It can be seen that although the shape can be adjusted so that the distribution of liquid indicates the relative radiation in any direction, the overall volume of liquid and hence the total radiation, remains the same.

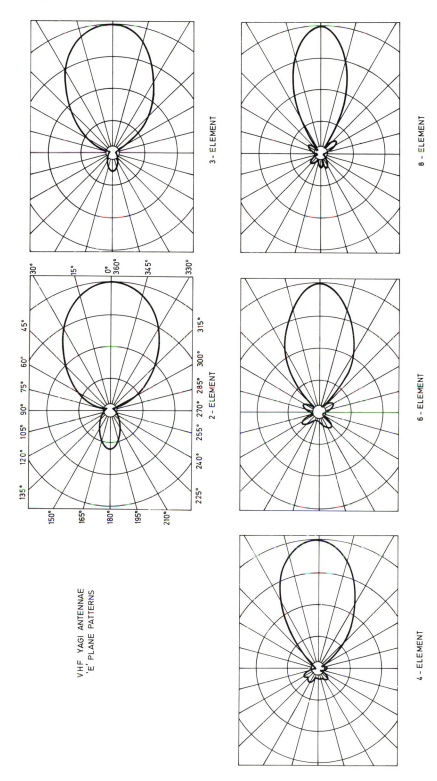

Fig. A23.2 Radiation Patterns of Various VHF Yagi Antennae (E-Plane)

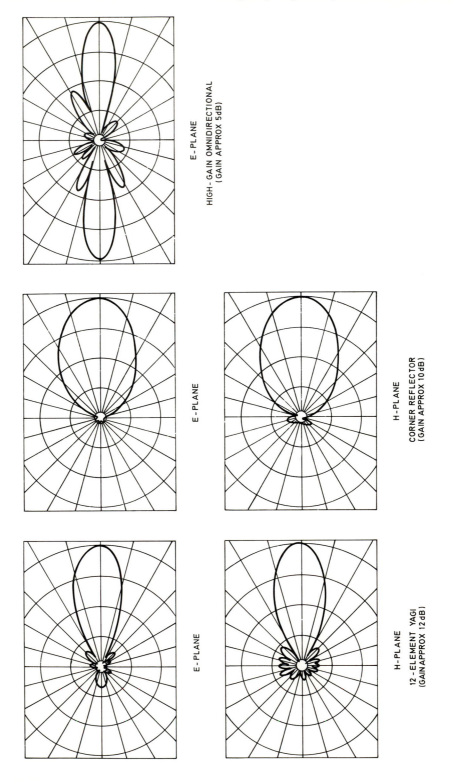

E - PLANE

HIGH - GAIN OMNIDIRECTIONAL
(GAIN APPROX 5dB)

E - PLANE

H - PLANE

CORNER REFLECTOR
(GAIN APPROX 10 dB)

E - PLANE

H - PLANE

12 - ELEMENT YAGI
(GAIN APPROX 12 dB)

Fig. A23.3 Radiation Patterns of Typical UHF Antennae

3. Antennae at Less than Free Space Conditions

If a half-wave dipole antenna is near to a reflecting object, any energy received by this object will obviously be reflected and will modify the effective shape of the radiation pattern of the dipole when measured at the far field. Further, if this reflecting object is adjusted in length to be resonant at the frequency in question, then the reflected energy will be at its maximum. Adjustment of the position of the object, so that it is in the same plane as the dipole at a spacing where the reflected energy is in the correct phase, will reinforce the main radiation in a particular direction. This is the first step in the development of a directional antenna.

From this it can be seen that if a dipole is supported by a metallic structure and this structure is sufficiently close to the dipole, its effect on the radiation pattern can be similar to that given above.

4. Effect of Mast Supporting Structure on a Half-Wave Dipole

4.1 Effect on Antenna Impedance. Before considering the effect of a nearby or supporting structure on the radiation pattern of an antenna, it is important to observe the effect of spacing on the impedance and the ultimate standing wave ratio obtainable.

Assume that a half-wave dipole in free space has its normal impedance of approximately 70 ohms. If this antenna is now placed in close proximity to a metallic structure with surface dimensions at least one wavelength in any direction, then the impedance of the antenna will change as the spacing between the antenna and the surface of the structure varies.

The approximate impedance resulting from the variations in spacing will be as follows:

Spacing from Structure	Impedance
0·5 Wavelength ($\lambda/2$)	70 ohms
0·33 Wavelength ($\lambda/3$)	100 ohms
0·25 Wavelength ($\lambda/4$)	83 ohms
0·16 Wavelength ($\lambda/6$)	50 ohms
0·1 Wavelength ($\lambda/10$)	20 ohms

Converting these values to effective VSWR, assuming a cable of 70 ohms characteristic impedance, we have:

Spacing from Structure	VSWR
0·5 Wavelength ($\lambda/2$)	1 : 1
0·33 Wavelength ($\lambda/3$)	1·4 : 1
0·25 Wavelength ($\lambda/4$)	1·2 : 1
0·16 Wavelength ($\lambda/6$)	1·4 : 1
0·1 Wavelength ($\lambda/10$)	3·5 : 1

Note:

$$\text{VSWR} = \frac{Z1}{Z2}$$

Assuming the use of feeder cable having an impedance of 50 ohms – this value of characteristic impedance is now generally accepted throughout the world – then the VSWR values will now become:

Spacing from Structure	VSWR
0·5 Wavelength ($\lambda/2$)	1·4 : 1
0·33 Wavelength ($\lambda/3$)	2 : 1
0·25 Wavelength ($\lambda/4$)	1·66: 1
0·16 Wavelength ($\lambda/6$)	1 : 1
0·1 Wavelength ($\lambda/10$)	2·5 : 1

To assist in appreciating the dimensions involved, the following table shows the various fractions of a wavelength relative to different frequencies.

Frequency	$\lambda/10$	$\lambda/6$	$\lambda/4$	$\lambda/3$	$\lambda/2$	λ
450 MHz	6·7 cm	11·1 cm	16·7 cm	22·3 cm	33·5 cm	66·7 cm
150 MHz	20 cm	33·4 cm	50 cm	66·7 cm	100 cm	200 cm
75 MHz	40 cm	66·7 cm	100 cm	133·3 cm	200 cm	400 cm
40 MHz	75 cm	125 cm	187·5 cm	250 cm	375 cm	750 cm

4.2   General Effect on Radiation Pattern. Fig. A23.4 shows typical mast cross-sections and associated antenna configurations. It can be seen that the supporting structure can vary from a single small diameter pole through various mast sizes up to a solid structure having dimensions extending for many wavelengths.

Obviously, the larger the mass behind the dipole the greater the tendency to limit the radiation pattern in the rearward direction. This particularly applies where the metallic structure is solid or with wire meshing where the mesh diagonal measurement is considerably less than a quarter wavelength. In the forward direction, the presence of a larger screen behind the antenna tends to reduce the width of the lobes produced, particularly those in the same plane as the structure.

4.3   Effects on Radiation Pattern as a Function of Spacing from Structure. The most important effect on a half-wave dipole is the change of its radiation pattern as the distance between the dipole and the structure varies.

If one considers the structure as a reflecting surface (referring back to Section 3 above), it can be seen that the spacing will affect the phase difference between the originating source in the dipole and the signal arriving back at the same point in the dipole after reflection.

The reinforcement pattern of the radiation will, therefore, vary and cause changes in the lobe pattern.

Examination of the first row of diagrams in Fig. A23.5 shows the change in radiation pattern as the spacing between a dipole and its supporting structure of $\lambda/8$ diameter varies. The broken concentric circles represent the radiation pattern of the antenna in free space. Subsequent rows of diagrams indicate the modification to these patterns made by increasing the effective structure diameter.

Further increases in the size of a solid structure reduce the radiation in the rearward direction still more and, ultimately, a structure size is reached where the radiation will not extend appreciably behind the structure.

With open structures, such as lattice masts and towers, the cut-off of the rearward radiation is defined but shows multiple lobes of narrow beamwidth and considerable depth. Fig. A23.6 indicates typical horizontal patterns of omni-directional antennae mounted on large cross-section towers. (a), (b) and (c) shows the effect of side mounting at various spacings, (d) the result of corner mounting

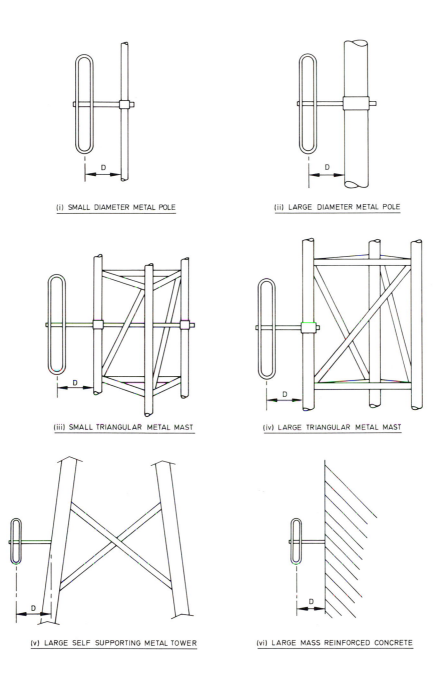

Fig. A23.4 Some Mast/Antenna Arrangements (D = Critical Spacing)

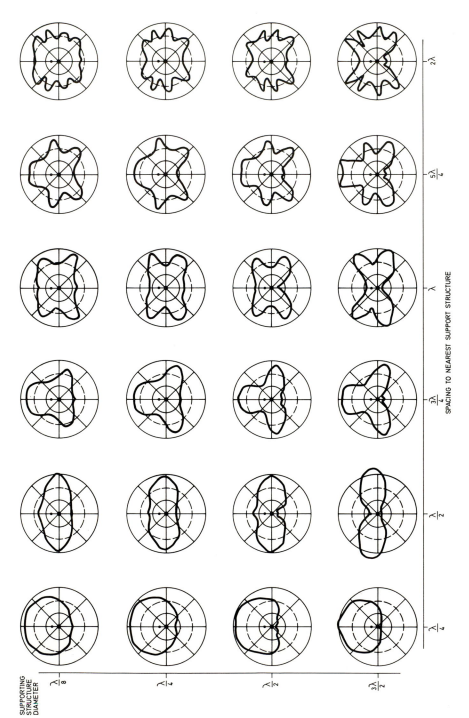

Fig. A23.5 Effect of Spacing and Structure Size on Radiation Pattern

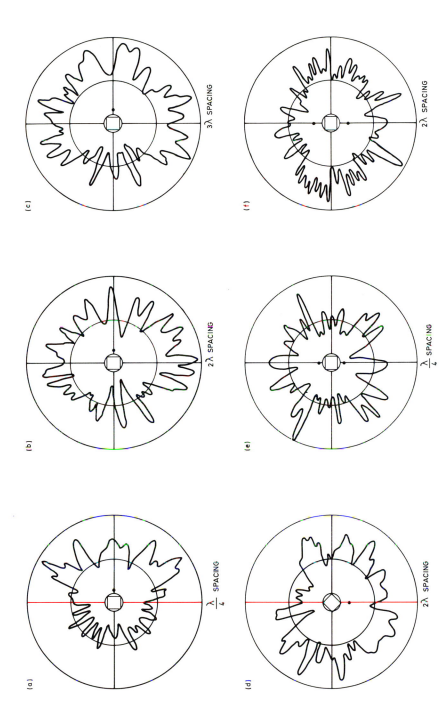

Fig. A23.6 Typical H-plane Patterns of Omni-directional Antennae Mounted on Metal Towers of Large Cross-Section (>5λsquare) for Various Antenna to Tower Spacings and (e), (f) the effect of mounting antennae on opposite sides of the tower and feeding them in phase.

4.4 Number of Lobes Produced. Examination of the patterns obtained (Fig. A23.5) highlights certain details. At each odd quarter wavelength spacing a forward lobe is produced, whilst for each even quarter wavelength two additional lobes appear, one in each direction in the same general plane as in the structure. Further analysis based on the above shows, therefore, that the resulting number of lobes is exactly equal to the number of quarter wavelength spacing intervals between the antenna and its support structure.

Where the number of vertical tower members extend over a considerable area behind the antenna, the number of lobes cannot be easily predicted.

4.5 Construction of Supporting Structure. The variation of the radiation pattern as a function of the structure details is of some importance. With an open form of construction there will be multiple reflections and, consequently, differing phase arrival times between the various reflections. This effect will tend to produce subsidiary lobes within the main pattern. Typical examples are shown in Fig. A23.6.

Further variations can be caused by a lossy structure and the mounting of an antenna on, say, the wall of non-reinforced building could well show considerable differences in radiation pattern under differing climatic conditions. Fortunately, the type of antenna normally mounted in this manner is usually directional and, therefore, a degree of protection exists at the rear of such an antenna by virtue of the rear reflecting element in the case of a Yagi or the screen assembly in the case of a corner reflector type. With such antennae, therefore, the variation with differing mounting structures can be less marked although possibly still objectionable under some circumstances. Where the effect is unacceptable, the inclusion of a mesh screen mounted on the face of the structure behind the antenna is to be recommended. Such a screen should not be less than a wavelength square.

The provision of a screen behind single or stacked or bayed Yagi antennae is often considered advantageous where minimum back radiation is required and where maximum isolation is required between antennae on opposite faces of a mast (see Fig. A23.7). The use of such a screen behind stacked or bayed antennae also serves as a more consistent form of construction with electrical constants less likely to cause variations in the overall pattern. It minimises effects caused by nearby antennae and metalwork. Again, the size of such a screen should not be less than a wavelength square per antenna in the array.

This type of construction explains some of the more consistent results obtainable with corner reflector antennae.

5. Use of Supporting Structure to Obtain Desired Radiation Pattern

As can be seen from analysis of Fig. A23.5, the desired patterns around a swing of not less than 180° can, theoretically, be obtained by suitably mounting the antenna at an optimum distance from the structure. However, it can be seen from Fig. A23.6 that structures with large open areas of lattice construction cannot be relied upon to produce "clean" radiation patterns with smooth outlines.

Some care must be exercised, therefore, when considering the use of the mast as a passive element.

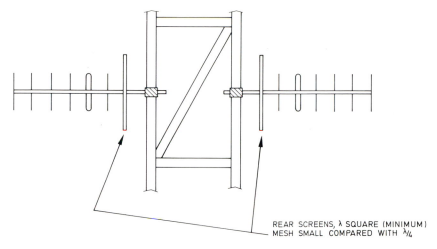

REAR SCREENS, λ SQUARE (MINIMUM)
MESH SMALL COMPARED WITH λ/4

Fig. A23.7 Using Back Screens to Minimise the Effect of the Mast and Coupling to
Other Antennae

The standard circular pole type of support provides similar pattern modification
to that of a conventional reflector element in a directional antenna and, therefore,
no difficulty should be experienced in providing the desired pattern.

With lattice masts, even those of small size, care should be taken when
positioning the antenna. Where possible, there should be one structure element
governing the position and other parts of the mast should be located symmetrically
behind the predominant structure element and screened by the main element as
much as possible. This infers that the mounting of an antenna on the corner support
will give the best results. Fig. A23.8 shows the possibilities. The dimension D should
be a multiple of quarter wavelength.

If possible, a mast joint should be avoided within a half wavelength of the
antenna centre. The reason for this is the possibility of non-linear rectifying action
producing spurious signals and intermodulation. Obviously the placing of a
rectifying joint in a radiation system can only lead to problems of mutual interference on the site and in its vicinity.

Where joints cannot be avoided and, indeed, even if they fall some distance
outside the main field, the use of copper straps by-passing them should be
considered during installation.

6. Omnidirectional Coverage from Side Mounted Antennae on Large Metallic Structures

Occasions may arise when omnidirectional coverage is needed from side mounted
antennae on large objects, such as water towers. Owing to the large area of screening
there will obviously be a lack of signal at the rear of a single antenna. The solution,
therefore, is to fill in such poor areas by providing as near an omnidirectional
radiation pattern as possible.

To achieve this, additional radiating antennae are required at strategic points on
the sides of the water tower. Referring to Fig. A23.9, it can be seen that each
individual antenna mounted at, say, a quarter wavelength from the structure would,

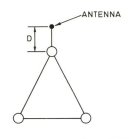

BEST POSITION ON TRIANGULAR
MAST. SYMMETRICAL RELATIVE
TO MAST CORNER.

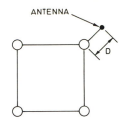

BEST POSITION ON SQUARE
MAST OR TOWER. SYMMETRICAL
RELATIVE TO MAST CORNER.

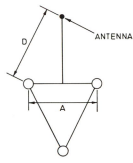

ALTERNATIVE POSITION FOR ANTENNA ON TRIANGULAR
MAST GIVING SLIGHTLY REDUCED LOBE WIDTH. ONLY
EFFECTIVE WHEN 'A' IS SMALL COMPARED WITH $\lambda/4$
AND 'D' = $\lambda/4$ (IF 'D' IS N x $\lambda/4$ THEN 'A' CAN BE
PROPORTIONATELY LARGER.)

Fig. A23.8 Minimising Unwanted Effects Caused by Structure

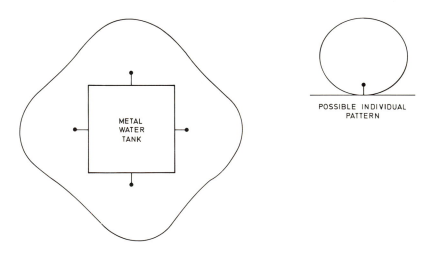

METAL
WATER
TANK

POSSIBLE INDIVIDUAL
PATTERN

Fig. A23.9 Antennae on Large Structure giving Omni-directional Coverage

if fed separately, provide substantially the pattern shown. To enable four such patterns to be additive and provide an all-round pattern, the feeding of the antennae must be such as to ensure correct phase relationships and for this reason the antenna array for a particular structure is usually designed for that sole purpose. The factors affecting such a design are complex and are not, therefore, included here. Furthermore, it would normally be necessary to perform a comprehensive field pattern test on the completed assembly to confirm its actual performance.

7. Conclusions

Whilst the foregoing notes indicate the general effect of a structure upon a given antenna, the patterns and ultimate performance of multiple antenna assemblies on large structures are often more complex. The radiation patterns shown in these notes are, in general, those obtainable with smaller structures; larger open structures (as shown in Fig A23.6) break up the patterns into lobes with much deeper nulls. Larger solid structures show a reduction in the number of lobes but generally at the expense of rearward radiation.

Appendix 24.   The Changes in Basic Characteristics Over the Frequency Spectrum From 30 to 1000 MHz

1.   General

Many of the shorter range radio services throughout the world operate in the 30 to 1000 MHz band. However, over this spread of spectrum, the characteristics vary considerably, some directly with frequency, some inversely with frequency whilst others are affected in accordance with other mathematical laws.

The following paragraphs, together with the associated tables, show these relationships.

2.   Noise Levels

The level of man-made noise expressed in dB above the thermal noise level at 288° K is shown in Table A24.1 (taken from C.C.I.R. Study Group Document DOC6/167-E (1972)). At the lower frequencies there is as much as 24 dB difference between the levels measured in rural and urban areas; at the higher frequencies the difference becomes insignificant.

Another comparison shows that the city noise at 450 MHz is approximately equal to rural noise at 75 MHz.

At 900 MHz, noise is virtually non-existent.

Galactic noise (from the solar system) approximates to a level between rural and quiet rural, whilst static noise (mainly caused by dry and thundery conditions) can be extremely high at the lower end of the spectrum. In the case of static, tropical areas provide the greatest levels.

Table A24.1 urban area noise is subject to some variation throughout the world depending upon local laws of noise suppression.

3.   Approximate Ranges

Table A24.2 shows approximate ranges over flat terrain between a 25W transmitter

FREQUENCY (MHz) CONDITIONS	30	50	66-88 (75)	150-174 (160)	450	900	COMMENTS
① CITY CENTRES (URBAN)	+36	+30	+25	+16	+4 ✳	0 ✳	MEAN VALUES OF MAN-MADE NOISE POWER FOR A SHORT VERTICAL LOSS-LESS GROUNDED MONOPOLE ANTENNA
② RESIDENTIAL (SUBURBAN)	+32	+26	+20	+11	0 ✳	0 ✳	
③ RURAL	+26	+20	+15	+6	0	0	LEVELS IN dB RELATIVE TO THERMAL NOISE AT 288° K
④ QUIET RURAL	+12	+6 ✳	+2 ✳	0 ✳	0 ✳	0 ✳	✳ EXTRAPOLATED
⑤ GALACTIC	+18	+13	+9	+1	0 ✳	0 ✳	
⑥ STATIC	HIGH	HIGH	MEDIUM	LOW	NIL	NIL	) ABSOLUTE FIGURES NOT AVAILABLE, BUT INVERSELY PROPORTIONAL TO FREQUENCY

Table A24.1 Noise Levels

FREQUENCY (MHz) CONDITIONS	30	50	66-88 (75)	150-174 (160)	450	900	COMMENTS
① k = 4/3 (PLANE EARTH & DIFFRACTION)	45 (28)	40 (25)	37 (23)	35·5 (22)	32 (20)	29 (18)	CONSTANT PATH ATTENUATION OF 150 dB
② k = 1 (PLANE EARTH & DIFFRACTION)					29 (18)	26 (16)	ASSUMING NO REFLECTION OR SCATTERING CONDITIONS EXIST
③ AS 1 BUT 100 FT, (30m) OBSTRUCTION AT CENTRE OF PATH	35·5 (22)	34 (19)	26 (16)	21 (13)	16 (10)	11·5 (7)	ALL RANGES IN KILOMETRES (MILES IN BRACKETS)
④ AS 2 BUT 100 FT, (30m) OBSTRUCTION AT CENTRE OF PATH					14·5 (9)	10 (6)	
⑤ AS 1 BUT WITH OVERALL AVERAGE INTERVENING BUILDING HEIGHT = 50 FT (15m)	10 (6)	6·4 (4)	5·7 (3·5)	4 (2·5)	3·2 (2)	1·6 (1)	VEHICLE IN STREET WITHIN 50 FEET (15m) OF GENERAL BUILDING LEVEL OF 50 FEET (15m)
⑥ DISTRIBUTING BANDS IN ACCORDANCE WITH THEIR LIKELY USAGE USING FIGURES FROM 1, 3 & 5	RURAL 45 (28) ①	RURAL 40 (25) ①	RURAL 37 (23) ①	SUBURBAN 21 (13) ③	URBAN/SUBURBAN 3·2 – 16 (2-10) (5 & 3)	URBAN 1·6 (1) ⑤	

Table A24.2 Approximate Ranges

feeding a dipole at a height of 100 ft to produce a level of 1 μV into a receiver with an antenna height of 5 ft. Note that, according to Bullington, from 40 to 450 MHz, the median level of signal at street level is 25 dB below the plane earth value. Over open terrain, the range difference appears to be less than 2 to 1 over the 30–1000 MHz spectrum.

Obstructions modify these approximate calculations. With a mid-path obstruction equal to the height of the fixed antenna, ranges could vary by 4 to 1.

In a city type of terrain, i.e. with an average building height over most of the area and normal intervening streets, the ranges are dramatically reduced and an approximate 6 to 1 range ratio is likely over the spectrum examined.

If we now allocate bands in accordance with their likely use, i.e. low bands to rural areas through to high frequencies for urban areas (based on the noise problem shown in Para 2 and Table A24.1), we can see a pattern which is more in keeping with that normally experienced and a range differential of up to 28 to 1 can be seen to apply.

4.   Transmitter Powers for constant signal to noise ratio

Reconsidering Table A24.1 and assuming urban area noise levels, it is obvious that *within the operational area*, RF powers must be increased at the lower frequencies to restore the signal to noise to a constant ratio.

If we take as a yard-stick 1 watt at 160 MHz, Table A24.3 (based on Table A24.1) shows, in the first example, the revised powers needed to restore the S/N ratio. Readjusting for propagation changes over a *non-obstructed path*, the second section shows the resulting outputs needed. The figures are only approximate and, if considered together with urban obstructions, etc., the tendency would be for the 450 and 900 MHz powers to be on the low side. The powers needed at these frequencies would be similar to those at 160 MHz.

CONDITIONS \ FREQUENCY (MHz)	30	50	66-88 (75)	150-174 (160)	450	900	COMMENTS
① CONSTANT WANTED INPUT SIGNAL INTO RX	100W	25W	8W	1W	65mW	25mW	BASED ON SIMILAR RECEIVER PERFORMANCE. FIGURES DO NOT TAKE INTO ACCOUNT RANGE ETC. AND ASSUME A MEASUREMENT LOCALITY WHERE BASIC S/N AT 160MHz IS SAY 30dB AND UNAFFECTED BY PROPAGATION CHANGES
② CORRECTED TX POWER ASSUMING FIXED TX-RX DISTANCE, I.E. INCLUDING CHANGES OVER AN UNOBSTRUCTED PROPAGATION PATH OF, SAY 10 MILES (16 kms) WITH H₁ = 100FT. (30m) H₂ = 5FT. (1·5m)	50W	16W	6W	1W	80mW	65mW	

Table A24.3 Transmitter Powers Required for Similar S/N Ratio (Relative to 1W) – Urban Area

5. Re-use Distances

These have been based on a previous paper in which a re-use distance of 5 times the operational radius of a station has been adopted.

Table A24.4 shows, in the first and second sections, the likely re-use distances over open terrain. If, however, we distribute the frequencies into the "likely use" categories of Table A24.2 Item 6, the re-use pattern follows general experience.

FREQUENCY (MHz) / CONDITIONS	30	50	65—58 (75)	150—174 (160)	450	900	COMMENTS
① TAKING 5x OPERATIONAL RADIUS EQUAL COVERAGE SYSTEMS K=4/3	225 (140)	200 (125)	185 (115)	177 (110)	160 (100)	145 (90)	RELATIVE TO RANGES GIVEN IN TABLE A24.2
② K=1					145 (90)	129 (80)	RANGES IN kM (MILES IN BRACKETS)
③ USING 5x OPERATIONAL RADIUS BUT ADJUSTING FOR RANGE UNDER NORMAL CONDITIONS FOR THE VARIOUS FREQUENCY BANDS (TABLE A24.2 ITEM 6)	225 (140)	200 (125)	185 (115)	104 (65)	16—90 (10—50)	8 (5)	

Table A24.4 Re-Use Distances (for similar heights)

6. Long Range Interference

As with noise, long range interference, whether caused by ionospheric changes, by temperature inversion or by a sunspot phenomenon, follows an inverse law with the lower frequencies being severely affected and the higher frequencies being generally immune from the effects. Table A24.5 shows the likelihood of long range interference in various circumstances.

FREQUENCY (MHz) / CONDITIONS	30	50	66—88 (75)	150—174 (160)	450	900MHz
① SPORADIC 'E' LAYER REFLECTION PRESENT	HIGH	HIGH	MEDIUM	UNLIKELY	NIL	NIL
② TEMPERATURE INVERSIONS ANOMALOUS PROPAGATION	HIGH	HIGH	HIGH	MEDIUM	UNLIKELY	NIL
③ SUNSPOT MAXIMA	HIGH	MEDIUM	UNLIKELY	NIL	NIL	NIL

Table A24.5 Likelihood of Long Range Interference

7. Antenna Wind Loading

The antenna size varies inversely with frequency and therefore each particular type of antenna will show markedly reduced wind loading as the frequency increases.

Above 900 MHz, however, when parabolic reflectors are used, wind loading can again become a problem particularly since the narrow beam can be noticeably deflected by the twist exerted on insufficiently supported structures. Table A24.6 shows typical figures.

FREQUENCY (MHz) CONDITIONS	30	50	66–88 (75)	150–174 (160)	450	900	COMMENTS
① FOLDED λ/2 DIPOLE (ANTENNA ONLY)	27·2 (60)	18·2 (40)	13·7 (30)	6·4 (14)	3·1 (6·9)	–	
② FOLDED GROUND PLANE	>32 (>70)	25 (55)	16·4 (36)	8·2 (18)	–	–	
③ HIGH GAIN MONOPOLE	–	–	8·2 (18)	6·1 (13·5)	3·2 (7)	–	FIBRE GLASS OUTER
④ CORNER REFLECTOR	–	–	–	–	44	–	GAIN ANTENNAE AT 900MHz GENERALLY PARABOLOIDS
⑤ YAGI 3 ELEMENT	–	–	33·2 (73)	16·8 (37)	6·8 (15)	–	
⑥ YAGI 6 ELEMENT	–	–	–	35 (77)	–	–	WIND LOADING DEPENDS ON SIZE AND CONSTRUCTION
⑦ YAGI 12 ELEMENT	–	–	–	–	20 (44)	–	WIND LOADING IN KILOGRAMS (POUNDS IN BRACKETS)

LOADING AT WIND SPEED 193·5km/h (120mph)

Table A24.6 Wind Loading for Half-wave Dipoles

FREQUENCY (MHz) CONDITIONS	30	50	66–88 (75)	150–174 (160)	450	900	COMMENTS
① MOBILE λ/4 WHIP	2 38 (7·8)	1·43 (4·7)	0·95 (3·1)	0·45 (1·5)	0·16 (0·52)	0·08 (0·26)	
② FIXED DIPOLE λ/2	4·76 (15·6)	2·86 (9·4)	1·9 (6·2)	0·9 (3)	0·32 (1·04)	0·16 (0·52)	
③ SPACING OF DIPOLE FROM MAST AT λ/4 (LENGTH OF MTG BOOM)	2·38 (7·8)	1·43 (4·7)	0·95 (3·1)	0·45 (1·5)			FURTHER INCREASES IN SIZE AND BOOM LENGTHS FOR GAIN ANTENNAE
④ SEPARATION (VERTICAL) BETWEEN 2 ANTENNAE BASED ON 30dB ISOLATION	8 5 (28)	5·5 (18)	3 (10)	1·8 (6)	0·9 (3)		
⑤ SEPARATION (HORIZONTAL) BETWEEN 2 (ANTENNAE) BASED ON 30dB ISOLATION	36·6 (120)	22·9 (75)	14 (46)	7·6 (25)	2·44 (8)		MEASUREMENT SHOWN IN METRES (FEET IN BRACKETS)
⑥ AVERAGE MINIMUM BEAMWIDTH WITH TYPICAL DIRECTIONAL ANTENNAE	POOR	POOR	POOR/AVERAGE	FAIR	GOOD	EXCELLENT	

Table A24.7 Approximate Antenna Sizes

8. Antenna Sizes

As mentioned in 7 above, the antenna size varies inversely with frequency and Table A24.7 shows the order of size, spacing, etc., likely with different antenna systems throughout the 30–1000 MHz spectrum.

9. Antenna Height

To be totally effective, to show correct relative gain with height increase and to exhibit the calculated gain and lobe shapes, antennae must be mounted at a minimum height above ground (water table).

Table A24.8 indicates the minimum height of antennae at various frequencies within the band 30–1000 MHz based on three types of ground surface – sea water, good conducting soil and poor soil (such as sand).

FREQUENCY (MHz) / CONDITIONS	30	50	66–55 (75)	150–174 (160)	450	900	COMMENTS
① SEA WATER (K=80 σ=4 MHO/m)	76 (250)	36·6 (120)	20 (65)	6 (20)	1·4 (4·5)	0·6 (2)	VERTICAL POLARISATION
② GOOD SOIL (K=30 σ=0·02 MHO/m)	8·8 (29)	5·5 (18)	3·4 (11)	1·5 (5)	0·6 (2)	≈0·3 (≈1)	HEIGHTS IN METRES (FEET IN BRACKETS)
③ POOR SOIL (K=4 σ=0·001 MHO/m)	4 (13)	2·3 (7·5)	1·5 (5)	0·7 (2·2)	≈0·3 (≈1)	≈0·3 (≈1)	

Table A24.8 Minimum Effective Antenna Height (Bullington)

10. Other Miscellaneous Factors

Those shown in Table A24.9 tend to be economic rather than electrical. Shown in 1 and 2 are the relative types of co-axial feeder cables and connector needed to provide approximately the same performance.

Obviously the better types can be used at the lower frequencies if circumstances dictate or permit.

The table also shows some figures from Bullington concerning tree and building path loss.

FREQUENCY (MHz) / CONDITIONS	30	50	66–88 (75)	150–174 (160)	450	900	COMMENTS
QUALITY OF COAXIAL FEEDER REQUIRED	HIGHER LOSS TYPES	HIGHER LOSS TYPES	MED. LOSS TYPES	MED. LOSS TYPES	LOW LOSS TYPES	LOW LOSS TYPES	LOSS (dB) INCREASES WITH LENGTH & $\sqrt{\frac{f2}{f1}}$ FOR ANY PARTICULAR CABLE TYPE
COAXIAL CONNECTORS (TYPICAL WORST CASE)	UHF S0239	UHF S0239	UHF S0239	UHF S0239 N TYPE	BNC } TYPES N	BNC } TYPES N	
TREE LOSS (BULLINGTON)	2–3 dB		4–9 dB			>1000 MHz OPTICAL LOSS = SOLID OBSTRUCTION	
BRICK WALL LOSS (BULLINGTON)	2–5 dB (DRY TO WET)					3000 MHz 10–40 dB (DRY TO WET)	

Table A24.9 Miscellaneous Factors

11. Summary

From the foregoing it can be seen that optimisation of the frequency to be used depends upon many factors and in broad terms really explains why frequencies between 65 and 174 MHz are usually chosen for conventional rural and urban systems with 450 MHz only being used for the normal urban type system.

Undoubtedly all frequencies have their merits; it is merely a question of assessing those merits against the broad background requirement and choosing the best compromise.

Appendix 25. Transmitter to Receiver Frequency Spacings with Relation to the Transmitter Noise Spectrum

1. General

With narrow transmitter to receiver spacings (i.e. those lying between the channel interval and 1 to 2 MHz), blocking characteristics of the receiver dictate the precautions which must be taken to avoid degradation of the wanted on-channel signals. Such precautions often include adequate physical spacing between transmitter and receiver antennae, feeders and equipments, and suitable R.F. filters to reduce the level of unwanted signals to an order which will not cause a signal to noise degradation by adversely affecting the receiver early stages. In short, either physical separation, improved receiver input selectivity, or a combination of the two can usually provide the necessary protection.

On the other hand, although transmitter noise can be attenuated directly by physical separation between the transmitter and the receiver or receivers, it is totally unaffected by receiver selectivity, RF or IF. It is effectively "on channel" and can therefore be alleviated only by increasing the antenna spacing or by reducing the noise output from the transmitter (either generally or at the discrete frequency of each of the affected receivers). A later figure illustrates these possibilities.

To enable the degree of noise reduction to be calculated it is, of course, necessary to know the magnitude of noise radiated from representative equipments and the following paragraphs indicate typical levels derived from various sources.

2. Methods of defining noise levels

The main method used by many theoreticians is one in which noise is specified in dB per Hz of bandwidth below carrier power or, alternatively, below 1 watt or 1 mW.

Such a method has merit insofar as it ties all details back to common level. It also, in theory, enables a much lower level to be determined, although, in practice, the measurement is often normalised from a measurement taken over a wider bandwidth and therefore such a gain may be hypothetical only.

However, we are normally concerned more with the effect of noise on a communication circuit and, for this reason, it is often advantageous to refer to the noise in a speech channel. Such a channel can, in link systems, be 3000 Hz wide but, in mobile radiotelephone systems, 2500 Hz is often the more normal figure. If all parameters in a system use the same basic bandwidth, they are easily related.

The following details are based on the second method.

3. Information Sources

Buesing (Ref. 5) states that, at frequencies very close to the carrier, the noise level approximates to 45 dB below carrier power and improves to about 80 dB at frequencies 15 kHz from carrier. He emphasises that modern technology is unable to attenuate further this level of "close in" transmitted noise spectrum.

This statement by Buesing therefore defines reasonably clearly and accurately the basic "close in" cut-off slope.

Beyond 15 kHz, equipment design determines the rate of attenuation. Bearing in mind that the level of noise is often, as a first approximation, a function of oscillator purity, it follows that as the number of tuned circuits following an oscillator increases, so will the cut-off rate. Multiplier stages do, however, introduce some further noise, which will, in turn, tend to lessen the improvement.

In "Spectrum Engineering – The Key to Progress" (Ref. 12) Section 6, graphs show the noise spectrum cut-off expected at 40, 150 and 470 MHz and Fig. A25.1

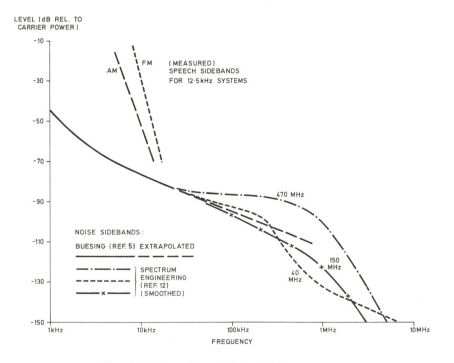

Fig. A25.1 Speech and Noise Sideband Spectrum

shows these curves smoothed to illustrate average conditions. An extrapolation of the Buesing figures is given, showing its general agreement with the Spectrum Engineering figures in the range 20 kHz to 2 MHz.

It will be seen that the smoothed cut-off curves of Fig. A25.1 tend to confirm the above points concerning the degree of cut-off versus tuned circuit attenuation.

4. Relation of speech and noise sideband spectrum

Fig. A25.2 shows an averaged cut-off together with the speech sidebands of typical AM and FM equipments (12·5 kHz). It shows quite clearly that, in the adjacent channel, the speech sidebands are the limiting factor but, two or more channels away, the noise sidebands take control and in fact up to a megahertz or so, only fall away at the rate of about 4 dB per octave.

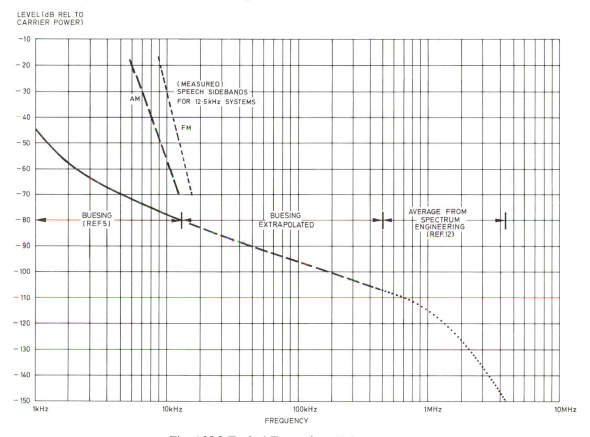

Fig. A25.2 Typical Transmitter Noise Spectrum

5. Protection required

The noise power in an ideal receiver will equal KTB watts where

    K = Boltzmann's constant ($1\cdot38 \times 10^{-23}$ Joules/Kelvin)
    T = Absolute temperature, usually 290° K
    B = Bandwidth (Hz)

Inserting a total bandwidth of 5000 Hz in the equation, the noise level becomes –167 dB.watt.

However, the normal receiver is not ideal and therefore some worsening of this figure can be expected. If we now consider a typical noise factor of 6 dB the level therefore becomes –161 dB.watt. With a wanted input signal of say, –149 dB.watt, it can be seen that the signal/noise ratio obtained will equal 12 dB.

We now imagine an unwanted signal, some kilohertz away, generating unwanted noise as described previously. If the level of this noise *in the received channel* is equal to –161 dB.watt, the noise level will rise from –161 to –158 dB.watt. (Note noise is non-coherent and therefore adds on a power basis). We therefore have a reduction or deterioration in signal/noise ratio of 3 dB down to 9 dB from the original 12 dB.

This reduction of 3 dB is a convenient level on which to base calculations, although Buesing and others tend to favour a 6 dB loss. However, for a 6 dB degradation, the added noise would be higher, equal to –156 dB.watt.

As we expressed the noise levels in dB.watt, we can continue and express the unwanted carrier power also in dB.watt. Then by using the average cut-off curve of Fig. A25.2 and locating its *relative* carrier level correctly, the point where the unwanted transmitter noise equals the receiver noise will show the frequency separation needed between the two units for a 3 dB degradation in signal to noise performance.

Fig. A25.3 shows an example where the transmitter power is + 10 dB.watt and the transmitter to receiver antenna isolation (attenuation between Tx and Rx antennae) is 35 dB. The *relative* carrier level thus equals + 10 – 35 dB (–25 dB.watt) and the transmitted noise at the receiver will be –161 dB.watt at 3 MHz from the transmitter frequency. Thus, with these figures, the frequency spacing should not be less than 3 MHz.

It can be seen that if the cut-off curve is drawn on a transparent loose cursor, it can be located relative to the *effective* carrier power and the frequency spacing read off directly. For clarity, as the main portion of the required cut-off is normally beyond a few hundred kilohertz from the transmitter carrier, it is preferable to express the frequency scale in a linear dimension as in Fig. A25.3.

6. Extra protection

If the specified frequency spacing is less than that shown as necessary in 5 above, additional protection will be required. The amount can be determined by use of the cut-off curves described in 5 and the cursor movement over Fig. A25.3 will indicate the figure quite simply.

In the example given, 3 MHz spacing is considered necessary with an effective transmitter carrier power of –25 dB.watt. If 1·2 MHz has been specified as a requirement, then, to achieve this smaller spacing, the *effective* carrier power must be reduced to –44 dB watt – a reduction of 19 dB. This is therefore the extra protection needed.

Three solutions are possible:

(i)   An increase in antenna isolation of at least 19 dB by increasing the spacing between the two antennae and/or by suitable rearrangement to maximise the attenuation using the E & H planes.

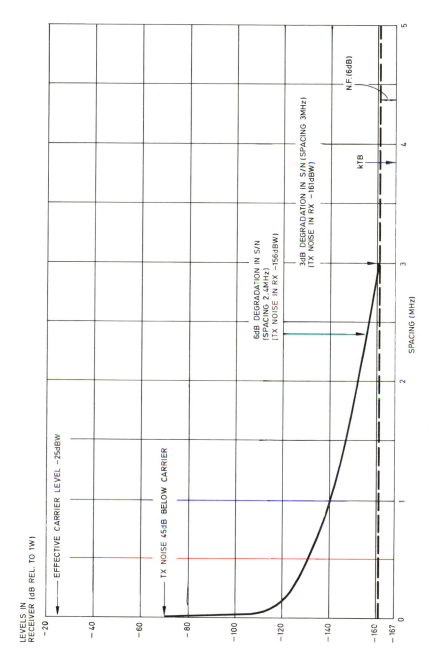

Fig. A25.3 Transmitter to Receiver Frequency Spacing and its Effect on Receiver Signal to Noise

(ii)   Adding a band stop filter in the transmitter output feed, with the resonant frequency adjusted to the receiver frequency. An attenuation in excess of 19 dB at that frequency and little or no loss at the transmitter carrier frequency will be needed.

(iii) Improving the transmitter output selectivity with a high Q bandpass filter tuned to the transmitter frequency. The cut-off must be such as to provide at least 19 dB extra attenuation at the frequency of the receiver nearest to the transmitter channel.

Equally, a combination of (i) and (ii) or (iii) giving a total of at least 19 dB attenuation would also provide the solution.

Where the transmitter is affecting several receivers on a number of nearby frequencies then (i) will obviously be relatively easy and cheap to implement, resulting in less transmitter output power loss and simpler maintenance. On the other hand, where space is at a premium, (iii) will probably be preferable if the power loss of the filter can be tolerated. Fig. A25.4 shows the various methods which can be used.

7. Conclusions

The degradation in performance of receivers operating in close proximity to transmitters radiating on nearby channels is not only caused by the blocking of those receivers by the high signal levels but also by the noise spectrum radiated by transmitters and causing an increase in residual noise in the affected receivers.

Blocking has been shown, in Appendices 6, 7, and 12 to be an effect which can be eliminated not only by transmitter to receiver isolation but also by suitable precautions at the receiver or receivers affected.

However, transmitted noise cannot be eliminated by any receiver precaution, the noise being intercepted by the tuned circuits within the receiver as an on-channel "signal".

Thus only transmitter to receiver isolation or removal of the noise at frequencies occupied by the receiver or receivers can be employed.

The graphs, obtained from available information, indicate the relationship between the required protection and the necessary frequency spacing.

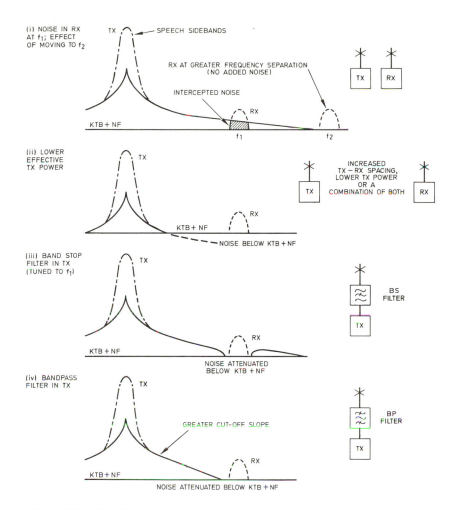

Fig. A25.4 The Effect of Filtering and/or Transmitter to Receiver Spacing

Appendix 26.   Choice of the Optimum Transmitter to Receiver Frequency Spacing as a Function of the Intermediate Frequency and Maximum Block Size

1.   General

The frequency spacing necessary between a single transmitter close to a single receiver has always been easy to define from the mode of operation. With simplex operation, very few problems exist, apart, for example, from ensuring that the antenna has sufficient bandwidth to cover the frequencies involved; with single frequency simplex, even this restriction does not exist. On the other hand, with duplex, equipment parameters such as blocking, intermodulation and transmitter noise have to be considered. A minimum frequency spacing can normally be defined for adequate isolation under average operating conditions.

There are also critical frequency spacings between a transmitter and a receiver operating in the duplex mode. These are based on the relation of the first intermediate frequency (or, to a lesser degree, a fraction of that frequency) to the transmit to receive spacing. Therefore, in a common spacing policy, such a possibility must be borne in mind.

It is common practice to allocate blocks of frequencies, suitably spaced as outlined above. However, as the number of equipments on a common site increases, so the probability of interference products also increases. This Appendix highlights those shortcomings which are directly attributed to the block size and spacing.

2.   Intermediate frequency

Many manufacturers currently use 10·7 MHz for the first intermediate frequency of most VHF and UHF receivers.* Where the transmit to receive spacing equals the IF, a high level of interference can occur. Furthermore, if the spacing equals IF/2 (5·35 MHz in this case), interference can also be sufficient to affect the system performance.

Any transmit to receive difference which falls at 5·35 or 10·7 MHz must therefore be avoided, including a small guard band of approximately ±50 kHz at 5·35 MHz and ±100 kHz at 10·7 MHz.†

So at any site, particularly sites on which a number of equipments are located, efforts must be made to avoid differences of 5·3 to 5·4 MHz and 10·6 to 10·8 MHz between *any* transmitter and *any* receiver on that site.

The interference, it should be noted, arises from the generation of 10·7 MHz, either directly, or, where 5·35 MHz is concerned, by a doubling action owing to non-linearity in the early receiver stages. The interference is caused by the transmitted signal at one or the other of the two spacings beating with the incoming wanted signal or even with the noise spectrum in the absence of a wanted signal. On communal sites, the interfering transmitter, even after taking extreme precautions, will generally still be sufficiently strong to provide an adequate mixing level and produce interference.

3.   Block size

Bearing in mind the frequencies to be avoided (see above), it is essential, for *full* protection, that the block sizes allocated to the individual transmitter and receiver segments must be a function of the transmit to receive spacing. There can, however, be some latitude if due consideration is taken when allocating individual channels.

Referring to Fig. A26.1, let us assume we have two equal blocks, one for transmitters and the other for receivers.

From the figure it can be seen that

$$f_2 - f_1 = f_{12} - f_{11}$$

Hence

$$f_{11} - f_1 = f_{12} - f_2$$

---

*A few manufacturers also employ a first IF of 21·4 MHz. Thus IF will equal 21·4 MHz and IF/2, 10·7 MHz. Such a combination will react in a similar manner to that described, with IF/2 providing the possible interference product when a 10·7 MHz Tx to Rx spacing is observed.

†The guard bands quoted are based on the assumption that coupling effects between the equipments, feeders and antennae of the transmitters and receivers concerned are reasonably low. If there is inadequate isolation, guard bands must be extended. For example, with adjacent units (vertically stacked in a common cabinet), it may be necessary to avoid a 10·7 MHz difference by as much as 1 to 2 MHz to achieve sufficient protection, whereas, with transmitters and receivers physically separated in different cabinets, the guard band can be as narrow as quoted. The need for physical separation therefore emphasises the recommendation (Section 5) that a transmitter to receiver frequency spacing of the order of 8 MHz is to be preferred.

If the frequency limits given in the previous paragraph are to be adopted, we have

$$f_{11} \ - \ f_2 \ = \ 5{\cdot}4 \quad \text{(Upper limit)}$$

$$f_{12} \ - \ f_1 \ = \ 10{\cdot}6 \quad \text{(Bottom limit)}$$

$$(f_{11} \ - \ f_2) \ + \ (f_{12} \ - \ f_1) \ = \ 16$$

Alternatively, rearranging we have

$$(f_{11} \ - \ f_1) \ + \ (f_{12} \ - \ f_2) \ = \ 16$$

Therefore, as $f_{11} - f_1$ and $f_{12} - f_2$ are identical, the *optimum* spacing will equal 8 MHz.

Continuing, we now see that, with this optimum spacing, the block size will be at its maximum and will equal

$$\begin{aligned} & \ 8 \ - \ (f_{11} \ - \ f_2) \\ = \ & \ 8 \ - \ 5{\cdot}4 \ \text{(Upper limit)} \\ = \ & \ 2{\cdot}6 \ \text{MHz wide} \end{aligned}$$

As the transmit to receive spacing is altered from 8 MHz to other values above or below that figure, the block size will decrease if full protection is to be maintained and will fall to zero when the spacing is exactly equal to IF or IF/2. Fig. A26.2 shows the general effect. Obviously, the limit in the downward direction is also governed by other parameters, such as blocking and transmitter noise. In the upward direction, factors such as common antenna bandwidths will prevail. Thus there will be a practical limit dictating the amount by which the spacing can be reduced or extended outside the frequency shown in Fig. A26.2.

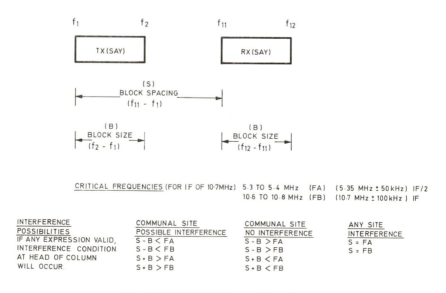

Fig. A26.1 Block Size and Spacing

4. Implications of extending block size or modifying Tx to Rx spacing

Although the optimum block size postulated above is obviously the ideal solution, there will be many occasions when either the block size will exceed that shown in Fig. A26.2 or when, owing to the block spacing, IF or IF/2 may fall within the blocks. Under these circumstances, on communal sites, at the best, some care and manipulation may be needed during the frequency planning stage; at the worst, conditions may become unworkable or require extreme filtering precautions. Between these two extremes there will exist conditions with increasing degrees of sterility involving parts of the blocks used; it will be necessary to exercise extreme care to avoid such slots.

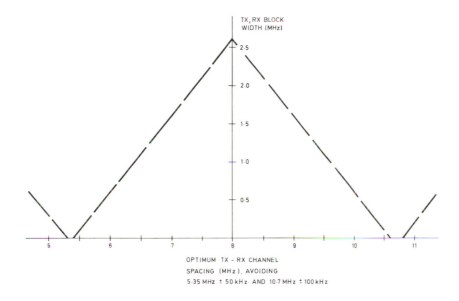

Fig. A26.2 Optimum Tx to Rx Channel Spacing as a Function of Block Width for Avoidance of IF Related Products

Let us, therefore, take typical examples of possible interference. Fig. A26.3 shows two blocks each of 3 MHz and separated by 4·6 MHz (a standard separation for many international systems) whilst Fig. A26.4 also uses 3 MHz blocks with the 10 MHz separation being recommended by the Conference of European Postal and Telecommunications Administrations (C.E.P.T.). Both figures indicate the possible interference potential.

Where, for example, a number of switched channels are required within a given system bandwidth, the probability of interference increases rapidly as the complexity of a communal site is extended.

If the allocated blocks, therefore, exceed the maximum size permissible to avoid IF/2 and IF interference or, alternatively, if IF/2 or IF can fall within the blocks as shown in Figs. A26.3 and A26.4, the frequency planning procedures for communal sites must observe the existence of "sterile" slots within those blocks.

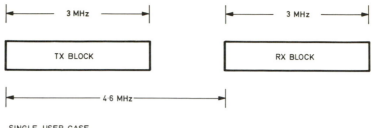

SINGLE USER CASE

TX = f_1 RX = f_2 SEPARATION = 4·6 MHz
NO INTERFERENCE

MULTI−USER CASE   (2 USERS, TYPICAL WORST CASE)

TXA = f_1 RXA = f_2 SEPARATION = 4·6 MHz
TXB = f_1 + 0·75 MHz   RXB = f_2 + 0·75 MHz   SEPARATION = 4·6 MHz
∴ SEPARATION   TXA(f_1) AND RXB (f_2 + 0·75 MHz) = 5·35 MHz = IF/2
INTERFERENCE CAN EXIST

CONCLUSIONS

1. WHERE FREQUENCY DIFFERENCE BETWEEN ANY TX AND ANY RX (ON A SITE HOUSING TWO OR MORE SETS OF EQUIPMENT) EQUALS 5·35 MHz (±50 kHz GUARD BAND), INTERFERENCE CAN OCCUR.

2. ON COMMUNAL SITES AS THE QUANTITY OF EQUIPMENTS INCREASES, THE PROBABILITY OF INTERFERENCE INCREASES RAPIDLY UNLESS SUITABLE PRECAUTIONS (SUCH AS FREQUENCY PLANNING CARE, RF FILTERING, etc.) ARE TAKEN.

Fig. A26.3   3MHz Blocks with 4.6MHz Separation

SINGLE USER CASE

TX = f_1 RX = f_2 SEPARATION = 10 MHz   NO INTERFERENCE

MULTI USER CASE   (2 USERS. TYPICAL WORST CASE)
TXA = f_1 RXA = f_2 SEPARATION = 10 MHz
TXB = f_1 + 0.7 MHz RXB = f_2 + 0.7 MHz   SEPARATION = 10 MHz
∴ SEPARATION TXA(f_1) AND RXB (f_2 + 0.7 MHz) = 10.7 MHz = IF
INTERFERENCE CAN OCCUR.

CONCLUSIONS

1. WHERE FREQUENCY DIFFERENCE BETWEEN ANY TX AND ANY RX (ON A SITE HOUSING TWO OR MORE SETS OF EQUIPMENT) EQUALS 10.7 MHz (±100 kHz GUARD BAND) INTERFERENCE CAN OCCUR.

2. ON COMMUNAL SITES AS THE QUANTITY OF EQUIPMENTS INCREASES, THE PROBABILITY OF INTERFERENCE INCREASES RAPIDLY UNLESS SUITABLE PRECAUTIONS (SUCH AS FREQUENCY PLANNING CARE, R.F. FILTERING, etc.) ARE TAKEN.

Fig. A26.4   3MHz Blocks with 10MHz Separation

Let us now examine this requirement:

> If S = nominal Tx to Rx spacing
> F = critical frequency (either IF or IF/2)

and $\triangle$ F = guard band requirement on either side of IF or IF/2, it can be shown that the *maximum sub-block size* will equal

$$S - (F + \triangle F) \text{ when S exceeds } F + \triangle F \tag{1}$$
$$\text{or} \quad (F - \triangle F) - S \text{ when } F - \triangle F \text{ exceeds S} \tag{2}$$

(1) will tend to apply in the IF/2 case with (2) generally arising in the IF case. It will, of course, be necessary to determine whether IF or IF/2 or both will prove troublesome and the diagnostic comments shown in Fig A26.1 should be used.

Similarly, it can be shown that the maximum width of the sterile slots in the blocks allocated must be

$$S - (F - \triangle F) \text{ when S exceeds } F - \triangle F \tag{3}$$
$$\text{or} \quad (F + \triangle F) - S \text{ when } F + \triangle F \text{ exceeds S} \tag{4}$$

Fig. A26.5 shows a typical example based on the use of 3 MHz blocks and a 4·6 MHz spacing, whilst Fig. A26.6 shows the condition arising in the case of a 10 MHz spacing with a 3 MHz block size. It must, of course, be emphasised that the position of sub-blocks (and sterile slots) in the total block can be varied between sites to occupy the blocks in the most efficient manner.

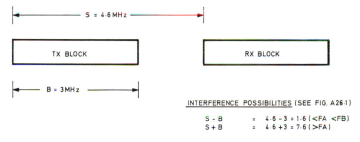

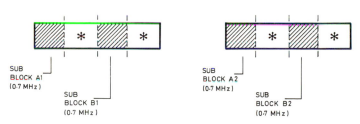

Fig. A26.5 Division of Sub-blocks on a Communal Site with 4.6MHz Transmit to Receive Spacing

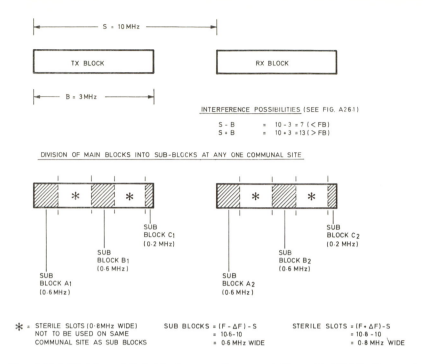

Fig. A26.6 Division of Sub-blocks on a Communal Site with 10MHz Transmit to Receive Spacing

To conclude, Fig. A26.7 (i) shows how the use of an optimum Tx to Rx spacing and its associated block size (taken from Fig. A26.2) results in a complete removal of sterile slots, whilst in Fig. A26.7 (ii) the effect of a Tx to Rx spacing very close to IF/2 (say 5·2 MHz) results in a huge array of sterile slots and very narrow useful sub-blocks. This second example is not a very elegant proposition if a number of closely spaced channels must be considered.

5. Submission

On the basis of the foregoing, it is obvious that, as the number of communal sites increases, there will be an extension of the problem which has already appeared in some of the more complex conurbations.

It is therefore submitted that the move to specify a common Tx to Rx spacing, a principle which exhibits considerable merit in standardising certain aspects of radio communication design, will, if not considered in relation to the potential interference prospects, be likely to cause increased frequency sterilisation at the more crowded communal sites.

The use of a Tx to Rx separation of the order of 10 MHz is certainly considered to be much too close to the 10·7 MHz intermediate frequency, currently in common use throughout the world. As this recommendation by C.E.P.T. to use such a separation will probably not be implemented until the further frequency plans have been discussed at the forthcoming World Administrative Radio Conference in 1979, it is strongly suggested that a more realistic spacing of around 8 MHz be adopted.

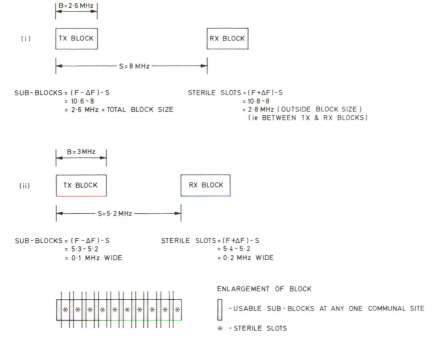

SUB-BLOCKS = (F − ΔF) − S
          = 10·6 − 8
          = 2·6 MHz = TOTAL BLOCK SIZE

STERILE SLOTS = (F + ΔF) − S
             = 10·8 − 8
             = 2·8 MHz (OUTSIDE BLOCK SIZE)
             (ie BETWEEN TX & RX BLOCKS)

SUB-BLOCKS = (F − ΔF) − S
          = 5·3 − 5·2
          = 0·1 MHz WIDE

STERILE SLOTS = (F + ΔF) − S
             = 5·4 − 5·2
             = 0·2 MHz WIDE

ENLARGEMENT OF BLOCK

⎮ — USABLE SUB-BLOCKS AT ANY ONE COMMUNAL SITE

✳ — STERILE SLOTS

Fig. A26.7 Comparison Between Tx-Rx Spacings and Blocks for Optimum Size, and
Tx-Rx Spacings Approaching IF/2 or IF

The frequency separation of 4·6 MHz, already in use in the VHF bands, suffers from a similar effect insofar as IF/2 is concerned. However, although IF/2 is illustrated as a possible hazard in this paper, its potential as an interference source must necessarily be less than that associated with IF and therefore some liberties can often be taken. Furthermore, the 10 MHz separation has been proposed by C.E.P.T. for use above 450 MHz, where it is obvious that the ability of the early stages of conventional receivers to reject signals 10 MHz from the nominal tuned frequency will be noticeably less than in the lower bands. The potential interference should be minimised by optimising the Tx to Rx spacing rather than by resorting to excessive and often loss-producing filter techniques.

Appendix 27. Definitions and Formulae

1. Channel Occupancy

1.1 Busy period. Normally a period of one hour (busy hour) out of the total 24 hour period during which the traffic passing over a given channel is at its highest level. It can, however, be a different length of time, if so desired, but it is normally accepted that one hour is the most convenient unit of time and is therefore usually adopted in most calculations.

1.2 Erlang. This is a traffic flow unit for a circuit or channel when continuously occupied for a specific base period. (For example, assuming one hour as the base period, full occupancy for the whole of that period would equal 1 Erlang. Similarly if 15 minutes be taken as the base period, full occupancy for 15 minutes would equal 1 Erlang and therefore 7½ minutes would then equal 0·5 Erlang.)

1.3 Average Holding Time (H) or Mean Effective Message Length (Me). The average duration in seconds, taken over an extended observation period, of the total time necessary to transmit a message, including all call establishment procedures, changeover periods, operational breaks, etc. Channel occupancy is based on this parameter.

1.4 Average actual message length. Average message information duration of a single user or group or similar users expressed in seconds and calculated from details derived from an extended period of observation.

1.5 Average number of Mobiles operating during the busy hour. This figure is derived from observations and indicates the average fraction of the total licensed number of mobiles operating during that period.

1.6 Average number of Mobiles operating per hour during the total 24 hour period. This figure indicates the average fraction of the total licensed mobiles operating during any one hour in the total 24 hour period. It is derived from observations of the circuit or channel.

1.7 Average number of calls per Mobile during the busy hour. This is derived from practical evidence and is the average number of calls each mobile operating during the busy hour is expected to make during that period.

1.8 Channel Occupancy per Mobile during the busy hour. This is the total period occupied by a mobile during the busy hour. It is the product of the average effective message length or holding time (3) and the average number of calls per mobile during the busy hour (7). The product as a fraction of the busy period expressed in seconds is the channel occupancy in Erlangs.

For example:

$$(30 \text{ sec. eff. message length} \times 2 \text{ calls})/3600 = 0.0167 \text{ Erlang}$$

Alternatively the reciprocal of the product can also be expressed as "mobiles per Erlang during the busy hour". Again, as an example, the above would equal 60 mobiles per Erlang in the busy hour.

1.9 Total channel occupancy during the busy hour. Total channel occupancy indicates the traffic flow and is derived from the product of channel occupancy per mobile during the busy hour (8) and the average number of mobiles operating during that period (5). This product gives the traffic flow in Erlangs (2) during the busy hour. It is often expressed as the percentage of full occupancy and the general formula applies:

$$\text{occupancy} = \frac{\text{t.v.c}}{36} \%, \quad \text{where:}$$

t = average effective message length in seconds.
v = average number of vehicles passing messages during the busy hour.
c = average number of messages per vehicle (call rate) during the busy hour.

1.10 Average Holding time to Average waiting time. This is a ratio dependent upon channel occupancy and follows an exponential curve based on

$$\frac{W}{H} \quad \text{or} \quad \frac{W}{Me} = \frac{T}{1 - T}$$

where T = traffic flow in erlangs.

For example, with an occupancy of 0.5 Erlang (50% of full occupancy) the W/H ratio = 1 indicating an average waiting time equal to an average effective message length or holding time in any random attempt to obtain channel access.

1.11 Average Effective message length to Average actual message length. This is an important measure of operating efficiency, namely the time wasted in *not* passing information.

Expressed as a ratio, average effective message length/average actual message length, obviously the more the ratio exceeds the theoretical best figure of 1, the lower is the circuit efficiency.

1.12 Deviation from average message length. The average message lengths (either effective or actual) are based on the average of a large number of messages of various lengths. There is therefore a chance that a given message will either exceed or be less than the average figure adopted for calculation. The extent of this deviation will determine the degree of reliability which can be attached to an average occupancy calculation.

It is also an indication of the spread of waiting to holding times likely to be calculated for a practical circuit.

1.13   Average number of calls per mobile over 24 hour period. This is an average *daily total rate* of messages passed by a mobile during a 24 hour period.

1.14   Figure of merit. The expression indicated in the previous paragraph (13) can provide the figure of merit of the circuit or channel when compared with the busy hour traffic.

The formula would be expressed as follows

$$\frac{\text{Average number of calls per mobile over 24 hour period (13)}}{24 \times \text{average number of calls per mobile during busy hour (7)}}$$

1.15   Grade of Service. This telephone term can be applied to mobile radio practice for comparison between channels with different loading characteristics. It is derived from the formula

$$\text{Grade of Service} \;=\; \frac{T}{1 + T}$$

where T is the traffic in Erlangs.

Thus with an occupancy of 0·5 Erlang (50% of full occupancy) the grade of service would be 0·33.

1.16   Lost Call Rate. The lost call rate is the reciprocal of the grade of service (15). It indicates the number of random calls which could be made for one call to be "lost", i.e. not possible or not established.

The formula would therefore be

One lost call in $(1 + T)/T$ attempts,

where T is the traffic in Erlangs.

Therefore, with the example given in (15), there would be one lost call in three random attempts to access the channel.

1.17   Loss of call value. This is an indication of the call value lost as a function of elapsed time. It is obviously dependent upon the type of message being passed. Plotkin of Harvard University, using a mathematical model, has derived an expression with an average effective length of message of around 30 seconds (Ref. 16).

$$\text{loss of value} \;=\; 100 \left(1 - e^{\left(\frac{-t}{20}\right)} \right) \%$$

where t is the delay in minutes.

As an example, if a delay of 5 minutes is assumed, it can be calculated that a 22% loss of value may occur.

The average value of t can also be considered as the waiting time H expressed in minutes and shown in (10).

2. Frequency Re-use

2.1 Re-use Factor. The number of times a frequency or channel is used within a given total area, for example, within a defined section of a country or within the country itself.

2.2 Re-use distance. The distance or distances between two or more sites on which the same frequency or channel is employed.

2.3 Coverage Radius. This is the operational radius from a given base station. It is the limiting radius defined by the minimum acceptable level of signal for satisfactory communications.

A simple approximation for this parameter over basically plane earth is shown by the formula (approximately true at 150 MHz):

radius in miles $= 1\cdot4 (\sqrt{h_1} + \sqrt{h_2})$

where h_1 = base station antenna height in feet

h_2 = mobile antenna height in feet

Alternatively

radius in kilometres $= 4\cdot1 (\sqrt{h_1} + \sqrt{h_2})$

where heights are in metres

2.4 Re-use distance based on coverage radius. For minimum interference under normal propagation conditions, using the formulae in (3), the re-use distance should approximate – using two frequency operation – to:

Distance between base station $= 4r_1 + r_2$

where r_1 = radius of coverage of the base station having the greater range
r_2 = radius of coverage of the base station having the smaller range

Alternatively, if both stations have similar ranges, the re-use distance should approximate to 5r where r is the operational range of either station.

3. Intermodulation

3.1 Definition. This phenomena is the generation of harmonically related spurious signals when two or more unwanted (or interfering) signals are combined in a non-linear device.

3.2 Even order intermodulation products. When two or more signals are involved in the generation of intermodulation, even order products can appear if the sum of the order of the harmonics is even. The products so generated are usually far removed from the bands of frequencies normally used for private mobile radio.

3.3 Odd Order intermodulation products. When two or more signals are involved in the generation of intermodulation, odd order products can appear if the sum of the orders of the harmonics is odd.

In this case two forms are generated, for example 2a + b and 2a – b. Only the latter fall in or near the bands involved, assuming the generating carriers are of the same order of frequency. The additive form will therefore be removed by a considerable amount from the bands involved.

A few examples of troublesome products falling in the wanted band are: 2a – b, a + b – c, 3a – 2b, 2a + b – 2c, 4a – 3b.

3.4   Order of products. The product order can be identified by the sum of the harmonic ratios of the various signals involved.
For example

> 2a – b = 3rd order
> 3a – 2b = 5th order
> a + b – c = 3rd order
> 2a + b – 2c = 5th order etc.,

3.5   Odd Order products generated from two signal sources. The number of odd order products of any one order falling in or near the wanted band i.e. 2a – b, 3a – 2b etc., can be calculated by the formula

> N (N – 1)

where N = number of channels available.

For example with 20 channels available there can exist

> 380   –   3rd order
> 380   –   5th order
> 380   –   7th order
> 380   –   subsequent odd order products

3.6   Third Order 2a – b products falling within a band of sequential adjacent channels. Where a number of sequential adjacent channels are allocated, the probability of 2a – b products generated from signals within this band and falling on wanted channels must be considered.

Two formulae indicate the number of products possible:

(*a*)   Even number of sequential adjacent channels

$$\text{Number of products} \;=\; \frac{N}{2}\,(N-2)$$

(*b*)   Odd number of sequential adjacent channels

$$\text{Number of products} \;=\; \frac{N}{2}\,(N-2) \;+\; 0\!\cdot\!5$$

In both cases N = number of sequential adjacent channels.

3.7   Third Order a + b – c products. The total quantity of possible products can be calculated by the formula

$$\left[\sum_{i=1}^{N-1} (i) \right] \quad (N-2)$$

where N = number of channels available.

3.8 Third Order a + b − c products falling within a band of sequential adjacent channels. As with the two signal case (6) two formulae indicate the number of products possible. The formulae are:

$$(a) \quad \left[\sum_{i=2}^{N-2} (i^2) \right] - \frac{(N-4)}{2} \qquad \text{Even number of sequential adjacent channels}$$

$$(b) \quad \left[\sum_{i=2}^{N-2} (i^2) \right] - \frac{(N-3)}{2} \qquad \text{Odd number of sequential adjacent channels}$$

28.　References

1.　INTERNATIONAL TELECOMMUNICATIONS UNION
"Table of Frequency Allocations 10 kHz to 40 GHz".
Geneva, 1966.

2.　BELLCHAMBERS, W.H., DURKIN, J., ROBERTSON, D.V.
"Computer-Assisted Frequency Assignment for the Private Land Mobile
Service".
Communications 74.
(Proceedings of conference organised by Wireless World and Electronics
Weekly.)
4.3. pp. 1–10.

3.　BRINKLEY, J.R.
"Land Mobile Radio, past, present and future".
I.E.E. Conference Publication No. 18, 1966, pp. 116–121.

4.　BRINKLEY, J.R.
"Mobile Radio Development".
British Communications and Electronics, July 1963, pp. 518–521.

5.　BUESING, Richard T.
"Modulation Methods and Channel Separation in the Land Mobile Service".
I.E.E.E. Transactions on Vehicular Technology, Vol. VT19, No. 2, May 1970,
pp. 187–206.

6.　BULLINGTON, K.
"Radio Propagation at Frequencies above 30 Megacycles".
Proc. I.R.E., Vol. 35, Oct. 1947, pp. 1122–1136.

7.　DAYMARSH, T.I., YUNG, T.J., VINCENT, W.R.
"A Study of Land Mobile Spectrum Utilization".
Part A – Acquisition, Analysis and Application of Spectrum Occupancy
Data, p. 149.
Stanford Research Institute, Menlo Park, California, U.S.A.
Project 7379, Contract RC10056, 1969.

8. EDWARDS, R., DURKIN, J., GREEN, D.H.
   "Selection of Intermodulation-free Frequencies for Multiple-channel Mobile Radio Systems".
   Proc. I.E.E., Vol. 116, No. 8, Aug. 1969, pp. 1311–1318.

9. HEWLETT PACKARD JOURNAL
   "Phase Modulation".
   Vol. 26, No. 11, July 1975, pp. 18–19.
   The Hewlett Packard Company, Palo Alto, L.A. 94304, U.S.A.

10. HOME OFFICE (U.K.) RADIO REGULATORY DIVISION
    Report on the Spectrum Requirements of Private Land Mobile Radio Services in the U.K., 1975–1999.
    Ref. MRC(75)4, May 1975.

11. ATKINSON, J.
    "Telephony", Vol. III, 1950, pp. 28–40.

12. JOINT TECHNICAL ADVISORY COMMITTEE OF THE I.E.E.E. AND E.I.A.
    "Urban Area Spectrum Usage. Spectrum Engineering – The Key to Progress".
    Section S5, March 1968.

13. MAGNUSKI, Henry
    "Improving Spectrum Utilization in Mobile Radio Communication".
    I.E.E.E. Vehicular Technical Group – 26th Annual Conference, Washington D.C., U.S.A., March 1976, pp. 105–110.

14. McCLURE, George
    "The Impact of System Design on 900 MHz Mobile Communications Service".
    Telecommunications, Dec. 1975, pp. 49–52.

15. MICROWAVE ASSOCIATES
    Leaflet No. 1074.

16. PLOTKIN, Howard A.
    "The Economic Comparison of One Way Paging and Full Two Way Radio Service".
    National Technical Information Service – U.S. Dept. of Commerce, Report No. COM-75-11141, April 1975.

17. PYE TELECOMMUNICATIONS LTD.
    "A Study of the Future Frequency Spectrum Requirements for Private Mobile Radio in the United Kingdom".
    Cambridge, England, 1976.

18. PANNELL, W.M.
    "The Case for 12·5 kHz Channels".
    Mobile Times, Dec. 1978, pp. 26–32.